Oil Fictions

Published in collaboration with the Society for Literature, Science, and the Arts, AnthropoScene presents books that examine relationships and points of intersection among the natural, biological, and applied sciences and the literary, visual, and performing arts. Books in the series promote new kinds of cross-disciplinary thinking arising from the idea that humans are changing the planet and its environments in radical and irreversible ways.

Oil Fictions

World Literature and Our Contemporary Petrosphere

Edited by
Stacey Balkan
and Swaralipi Nandi

The Pennsylvania
State University Press
University Park,
Pennsylvania

Chapter 12 is an edited and abridged version of "Conjectures on World Energy Literature: Or, What Is Petroculture?" by Imre Szeman, in the *Journal of Postcolonial Writing* 53, no. 3, published May 4, 2017, by Taylor and Francis. Reprinted by permission of Taylor and Francis Ltd, http://tandfonline.com.

Library of Congress Cataloging-in-Publication Data

Names: Balkan, Stacey, 1973– editor. | Nandi, Swaralipi, editor.
Title: Oil fictions : world literature and our contemporary petrosphere / edited by Stacey Balkan and Swaralipi Nandi.
Other titles: AnthropoScene.
Description: University Park, Pennsylvania : The Pennsylvania State University Press, [2021] | Series: AnthropoScene: the SLSA book series | Includes bibliographical references and index.
Summary: "Explores literature and film about petroleum as a genre of world literature, focusing on the ubiquity of oil as well as the cultural response to petroleum in postcolonial states"—Provided by publisher.
Identifiers: LCCN 2021023466 | ISBN 9780271091587 (hardback) | ISBN 9780271091594 (paper)
Subjects: LCSH: Petroleum in literature. | Fiction—20th century—History and criticism. | Fiction—21st century—History and criticism. | LCGFT: Essays.
Classification: LCC PN3352.P38 O45 2021 | DDC 809/.93356—dc23
LC record available at https://lccn.loc.gov/2021023466

Printed in the United States of America
Published by The Pennsylvania State University Press,
University Park, PA 16802-1003

The Pennsylvania State University Press is a member of the Association of University Presses.

It is the policy of The Pennsylvania State University Press to use acid-free paper. Publications on uncoated stock satisfy the minimum requirements of American National Standard for Information Sciences—Permanence of Paper for Printed Library Material, ANSI Z39.48–1992.

Contents

Jennifer Balkan, *Mortality*, 2018. Oil on panel, 36 × 24 inches.

Preface

Stacey Balkan and
Swaralipi Nandi

Oil Fictions is a transnational collaborative project, with its editors and contributors located in diverse global contexts and connected through metaphorical pipelines of an increasingly precarious environment. Having never met beyond the virtual world, we editors connected over a common concern for our shared living in the "real" world. *Oil Fictions* began to take shape in the fall of 2017 amid *unprecedented* environmental chaos. A series of potentially cataclysmic storms would form in the Atlantic Ocean—Irma, a Category 5 hurricane, caused the evacuation of Stacey's community in South Florida; and a sustained heat wave, hovering above 46 degrees Celsius (115°F), would plague the Indian state of Telangana where Swaralipi teaches and lives with her family. As we finished work on this volume, Hurricane Michael left unimaginable destruction in Florida, and India was recovering from a devastating flood in Kerala, the worst in a hundred years. Such catastrophic weather events have become commonplace in the era popularly referred to as the Anthropocene, but Telangana, Kerala, and South Florida offer uncannily concise indices of the deleterious effects of fossil capitalism, extreme extraction, and feckless development. Not to mention that the "MOUist corridor" in central Andhra (adjacent to Telangana), documented by Arundhati Roy in her 2011 *Walking with the Comrades* and labeled as such for the scores of MOUs (memoranda of understanding) on every "mountain top" and "blade of grass" in the region, precisely mirrors the legal land grabs sanctioned by private industry across South Florida. The result has been the decimation of local ecologies and the increased vulnerability of local communities to extreme weather events. As such, these diverse states figure as central coordinates in the global petrosphere that the present study seeks to map; so too does the possibility of collaboration and a spirited conviviality in these desperate times.

A distinct objective of *Oil Fictions* has been to explore literature on petroleum as world literature, focusing on the ubiquity of oil as well as the cultural response to petroleum in postcolonial states. Petrocultural discourse has largely been tethered to cultural production in the Global North. When we started conceptualizing this project, we were haunted by the glaring lacuna of a sustained petrocultural paradigm in postcolonial contexts—barring a couple of independent critical essays. It has therefore been a central aim of this volume to engage with petrofictions in a variety of postcolonial and world literature milieus: African, South American, South Asian, Middle Eastern, and transnational encounters addressing the all-pervasive oil economy. Another ambition of the volume has been to foreground the human cost of petroleum extraction while also grappling with the vexed categories of the "human" and "nature"—those real abstractions that have long sustained global capitalism. Our volume engages with literature that represents a trajectory of imperial power imbalance as well as the extractive labor regimes of the transnational petro-economy. The volume thus intervenes in the idea of petroleum as a given, inevitable aspect of modern life and instead investigates the complex power structures that sustain our global petrosphere.

Oil Fictions has also tried to strike a balance between the renowned scholars of the field and a host of emerging voices. When we floated the call for papers, we were overwhelmed with the response, not only in the volume of abstracts we received but also in the discovery of the myriad ways in which petroleum culture is theorized, imagined, and resisted through literature all over the world. Our volume has thus spilled over theoretical boundaries and genre categorizations; it investigates petrofiction as a genre of world literature and film, the relationship of colonialism to the fossil fuel economy, issues of gender through ruminations on petrofeminism in the Thermocene epoch, and discussions of migration, precarious labor, and the petro-diaspora. The volume's uniqueness is enhanced by the inclusion of testimonies of the oil encounter—through memoirs, journals, and interviews—from a diverse geopolitical grid ranging from the Permian Basin to the Persian Gulf. Indeed, the collection as a whole instantiates a joint effort to recognize the wide-reaching effects of fossil-fueled tyranny while also cultivating a transnational community of scholar-activists committed to fighting for social and environmental justice.

In the chapters that follow, we labor to recognize the material resonances of the aesthetic and the role of the imaginative in the "urgent envisionings" necessary to move beyond petroleum;[1] we chose the image on page viii to reflect just this. *Mortality* casts three petroleum-based plastic ribbons, vital and aglow, against the extinguished life of a single flower. The painting beseeches its viewers to consider the finitude of mortality. As petro-critics we understand the work as a call to recognize the imperium of artifice and of the commodity form and thus the imperative of our work to cultivate life in the ruins of capital.

Note

1. Shelley Streeby, *Imagining the Future of Climate Change: World-Making Through Science Fiction and Activism* (Oakland: University of California Press, 2018).

Acknowledgments

A project of this scope seemed quite daunting to me as a first-semester assistant professor starting the tenure track. My gratitude for the community of thinkers, scholars, teachers, activists, friends, and family who made this possible is profound. I must first thank my colleague, friend, and coeditor Swaralipi Nandi, with whom I hope to continue working toward energy justice. She has become an indispensable part of my community.

I also want to thank my colleagues and friends at Florida Atlantic University. As the first hire in Environmental Humanities, I was given the freedom to pursue a research program and cultivate a pedagogical practice that reflected my commitment to environmental and energy justice. I am grateful for my colleagues and for my students—particularly Eric Berlatsky, my department chair; Andrew Furman, my faculty mentor; and Devin Garofalo, my dear friend and brilliant interlocutor who read multiple drafts of several components of *Oil Fictions*. A warm thank you especially to Jake Henson and Ashvin Kini for their generous comments and consistent support.

My thinking about the stakes of petromodernity and the political virtue of petro-aesthetic expression was first sown in a paper for the 2015 convention of the Modern Language Association (MLA) on the petro-picaresque. The paper then blossomed into an article for a special issue of *The Global South*, generously edited by Leigh Anne Hunt and Sabine Haenni. It was also at a meeting of the MLA that I joined a panel discussion, convened by Anne Garland Mahler, regarding our contemporary petrosphere; the panel's namesake served as inspiration for the title of this collection. In this vein I also thank Corbin Hiday, who not only contributed to *Oil Fictions* but also convened a panel session at the annual meeting of the Northeastern Modern Language Association (NeMLA) on Marx and ecology where I first presented my chapter for the book.

My thinking about petrocultures began in earnest upon exploring the work of Amitav Ghosh, whose foundational essay on petrofictions ostensibly laid the groundwork for the field. Ghosh's work has also been a consistent presence in my thinking about postcolonial environments and environmental justice, and I am deeply grateful for his collaboration on and contribution to *Oil Fictions*. I am indebted also to my teacher and mentor Ashley Dawson at the CUNY Graduate Center, who taught me the meaning of the term *scholar-activist* and whose participation in this project has filled me with great joy and pride; the same is true of Micheal Angelo Rumore, my CUNY comrade, friend, and collaborator. Imre Szeman has been a true beacon and a mentor both on *Oil Fictions* and in the broader field of Petrocultures. His consistent support, including an invitation to participate in the second *After Oil* school in Montreal, has been critical.

It is to Swaralipi, Imre, and all of the contributors to *Oil Fictions* that I offer my sincerest thanks—for their spirited and tireless commitment to this project and so much more. As we come to the end of this long road, amid a pandemic that reminds us not just of the endemic political toxicity that plagues our global community but also of the potential for communal solidarity, my heart is full with the spirit of this bright collective.

My family and dear friends have also been a consistent source of support in the three-plus years that this project spanned. The call for papers for *Oil Fictions* was first conceived after my community in South Florida had been evacuated during Hurricane Irma in the fall of 2017 (three short weeks after I had moved to Florida). I thank Bonnie MacDougall and Don Grein for offering me shelter from the storm in more ways than I can enumerate—that I penned a draft of the CFP via generator seems particularly appropriate. Thank you, Bonnie and Don! I am grateful for my loving chum Christopher Bosek, who introduced me to the majestic hinterlands of my adoptive swamp, who brought me back to the woods, and who bestowed upon me my dear (nonhuman) research assistant, Asbury. My sister, Jennifer Balkan, continues to be my greatest inspiration—a true teacher, and my first teacher. It is with endless pleasure, admiration, and gratitude that Jennifer's painting *Mortality* graces this volume.

Much of what we do in this field is because we are committed to the welfare of future generations. It is in this spirit that I dedicate my work on this volume to my beloved nephew, Karlo Bloom.

Stacey Balkan
Delray Beach, Florida, United States
April 17, 2020

This book came at a crucial juncture in my postdoctoral research stasis, whereby looking for new directions to explore, I stumbled upon the field of Energy Humanities. My sincerest gratitude goes to my friend, colleague, and coeditor, Stacey Balkan, who believed in the project as soon as I proposed it (and after we had just "met" on academia.edu!) and took it to new heights with her enthusiasm and expertise. She has been the steering force whenever my pace slackened.

I thank my colleagues at Loyola Academy for their constant encouragement and support, for believing in me, and for their valuable suggestions and edits. More importantly, I thank them for being such supportive coworkers when I juggled this project with a full teaching load and the administrative duties of a departmental chair. My students keep inspiring me, and their emotional support has been invaluable during this journey. The fresh yet insightful perspectives of these young minds have been instrumental in thinking about the world in new ways.

My heartfelt gratitude goes to the members of the petrocultures collective—being part of this group has constantly sparked new ideas and opened new avenues. Their passion for and commitment to environmental justice gives hope for a more meaningful collective endeavor, regardless of our scattered global locales.

I extend special thanks to Professor Imre Szeman for his spectacular support of this project. As a pioneer in the field, he has obviously inspired us through his scholarship. But he went much beyond that, mentoring us throughout the journey and connecting with us personally. I hope to continue working with him and will look forward to his much-awaited India visit in the post-Covid world! A word of gratitude goes also to Professor Graeme Macdonald—another stalwart in the field—for his extremely encouraging review of the manuscript, which steered it to publication.

I thank all our contributors for their patience and commitment. It's been quite a long journey since the time we floated the CFP, and yet none of them gave up on us. Never once have they expressed their displeasure at the long wait or failed to attend to our revision requests promptly. Thank you, dear contributors, for believing in us and entrusting us with your valuable work.

Finally, I have immense gratitude for my amazing family, who define my existence—my parents and my parents-in-law, for reminding me of my potential and for never giving up on me, and my brother, whose love and admiration inspires me beyond words. And no amount of thanks is enough for the man who stands with me like a rock—my husband, Sayan—who is perhaps more excited about my projects than even I am! Most important, my gratitude goes to the one I love the most—my son, Shreyaan—who teaches me lessons in kindness, empathy, and caring every single day. I dedicate this volume to him and to all the young ones. I hope we leave a better world for them.

Swaralipi Nandi
Hyderabad, India
April 17, 2020

Introduction
Reading Our Contemporary Petrosphere

Stacey Balkan and
Swaralipi Nandi

In the day we sweat it out on the streets of a runaway American dream
At night we ride through the mansions of glory in suicide machines
Sprung from cages out on highway nine,
Chrome wheeled, fuel injected, and steppin' out over the line
Oh, Baby this town rips the bones from your back
It's a death trap, it's a suicide rap
We gotta get out while we're young
'Cause tramps like us, baby we were born to run
—Bruce Springsteen, "Born to Run" (1975)

If the spice trade has any twentieth-century equivalent, it can only be the oil industry.
—Amitav Ghosh, "Petrofiction" (1992)

Bruce Springsteen's "Born to Run" topped Billboard charts in the United States just two years after the OPEC (Organization of the Petroleum Exporting Countries) embargo crippled the US economy and drove oil prices well above the reach of middle-class wages.[1] However much Springsteen's lyrics critique the material forces that were destroying working-class communities in his native New Jersey, the song proved to be an anthem—a tribute to the "fuel-injected" American dream—whose release dovetailed neatly with the collapse of the union shop and the ascendancy of Reagan-era neoliberalism.

While the materiality of petroleum has proved stubbornly resistant to any serious cultural critique—evidence of what Rob Nixon has termed "spatial amnesia"—its ubiquity registers in such aesthetic productions.[2] So, too, critics of the industry, if also of the global apartheid system that constitutes our contemporary petrosphere, abound in literatures that span the so-called American century. Long before poet A. R. Ammons indicted American industry, dramatizing the visceral materiality of oil and sewage and calling for industrial waste to be "the poem of our time,"[3] Aldo Leopold condemned the nation's cultural myopia in his 1933 *Game Management*. "When I go birding in my Ford," he quipped, "I am devastating an oil field and re-electing an imperialist to get me rubber."[4] Writing just one year before the collapse of Henry Ford's disastrous Fordlandia—the Brazilian utopia where Ford would enclose native forest for rubber monoculture—Leopold beseeched his readers to consider the evident material substrate of their aesthetic pleasure.

Of course, such stories wouldn't resonate with the popular American imagination, committed as it was to a carbon-intensive middle-class existence made possible by the invisibility of such labor regimes as might be glimpsed in the oil fields of Nigeria or the rubber plantations of Brazil or Myanmar. Leopold's indictment would thus seem heretical to most, and perhaps even absurd: with new forms of carbon consumption making possible the Levittowns of the next decade and paving the way for the forthcoming interstate highway system, the nation's commitment to oil extraction was reaching a fever pitch. Despite the resonance of his remarks for an environmental consciousness forged over generations of accumulation by occupation and dispossession, America's landscape ethic—conceived in equal parts by the likes of Thomas Jefferson, Frederick Jackson Turner, and Henry David Thoreau—thrived on its ability to sublimate historical trauma to such aesthetic ideals as the neatly ordered fields of the plantation, the democratic vistas of the frontier, the "vacant" corners of wilderness in Thoreau's Katahdin, and ultimately the suburban expanses of the postwar era, with its V-8 engines and new modes of racial segregation. Surely, there was no room in such a morally bankrupt aesthetic tradition for labor, and certainly not for the ever-ephemeral oil field.[5]

In what we might call, following Fredric Jameson, the "present absence" of oil in the American imagination—essentially the constitutive outside that has long forged the colonial elsewheres of our imperial imaginary—oil has effectively resisted representation owing to both a conscious refusal to acknowledge the human and environmental costs of its extraction as well as the aforementioned spatial amnesia, which renders such sites invisible. The central focus of this study is in part the stark invisibility of energy systems, along with the

Indigenous communities removed in the interest of extractivist networks of accumulation that have long powered cities in the Global North. The essays collected in *Oil Fictions* critique the silent ubiquity of oil through readings of what, following cultural critic Imre Szeman, we call oil fiction—a variant of Amitav Ghosh's petrofiction in which the trope of the oil encounter is often sublimated within the broader lineaments of our contemporary petrosphere. By seeking to make visible a mode of "violence discounted by dominant structures of [imaginative, often visual, but also literary] apprehension"—that of extractivist economies that thrive on the off-siting of violence—*Oil Fictions* stages a critical intervention that aligns with the broader thrust of the Energy Humanities.[6]

We work under the premise that "energy is dialectically bound to economic history—not a concept or variable independent of it, but a structuring force without which capital could not operate"—and that energy, in the form of fossil capital, expresses itself aesthetically in ways that subvert dominant visual paradigms.[7] To the extent that modernity is in fact made possible by fossil fuels, we argue for an understanding of oil fiction that recognizes the necessary material conditions of its production as well as the always already intertwined phenomena of extractivist regimes and the dispossession of the poor and marginalized.

Following both Shelley Streeby's remarkable study of the aesthetics of protest in the era of fossil capitalism and work on what Brent Ryan Bellamy and Jeff Diamanti have described in terms of resource aesthetics, we do not distinguish between economic and aesthetic production.[8] Instead, we recognize and seek to make visible the evident materiality of petroculture. We consider, *pace* Patricia Yaeger, "what happens if we sort texts according to the energy sources that made them possible."[9] That is, we ask "how . . . might one best begin to use energy as the lens through which to view culture."[10] Karina Baptista explains: "Broadly defined . . . [petroculture] refers to the social imaginaries constituted by the knowledge, practices, and discourses resulting from the consumption of and subsequent dependence on oil"—essentially, the "carbon-heavy forms of middle-class American life" and attendant narratives of mobility and freedom of the sort captured in a Bruce Springsteen song.[11]

The dissonance between the destructive nature of petroleum—economically, politically, and ecologically—and the "abstract categories" that it enables, including freedom and mobility, is the central preoccupation of cultural critics working in the emergent field of Energy Humanities, in which the study of petroculture is central.[12] Based at the University of Alberta some seven hundred kilometers from the Alberta Tar Sands, the Petrocultures Research Cluster has distinguished a critical field to investigate the (broadly conceived) implications of petroleum in our "world ecology."[13] In their recent anthology, Imre Szeman and Dominic Boyer attend to our "energy unconscious" through a critique of the "fuel apparatus of modernity, which [is] all too often invisible or subterranean, but which pumps and seeps into the groundwaters of politics, culture, institutions, and knowledge in unexpected ways."[14] The authors also make clear that the Energy Humanities, as a project, "aspires to provide a speculative impulse as well as a critical diagnostics"—a means of world-making in an era of climate chaos.[15]

Tracing a genealogy of petrocultures and criticism, such critics have generally cited Amitav Ghosh's 1992 essay "Petrofiction" as a foundational text. In the essay, in part a review of Abdelrahman Munif's *Cities of Salt*, Ghosh asks, "Why, when there is so much to write about, has [the oil] encounter proved so imaginatively sterile?" In his 2016 monograph *The Great Derangement: Climate Change and the Unthinkable*, Ghosh returns to this central question in what he calls "an alternative history of the carbon economy." Here, as if recognizing the limitations of a "petrofiction" reducible to images of towering derricks, he remarks on the petrocultural phenomenon of the road novel in American letters, jesting that it is only a matter of time before a graduate student finally calculates the carbon emissions for which these novels are responsible.[16]

The project, however, is ultimately a lament of our "great derangement"—that is, the persistent dearth of fictional representations of the environmental calamity wrought by our feckless dependency on fossil fuels. In this sense, Ghosh's more recent query resonates with his earlier essay while also anticipating a sentiment issued by Szeman and Boyer, who remark, "If it has been so difficult to grasp and grapple with so important an element, it is in many respects because fossil fuels are saturated into every aspect of our social substance."[17] *Oil fiction*, in Szeman's formulation, thus replaces *petrofiction* as a term to indicate a more capacious understanding of the genre. In this sense, and *pace* Ghosh's argument, *On the Road* would join Upton Sinclair's *Oil!*—the latter, like *Cities of Salt*, a more conventional, because explicit, critique of the oil industry. The 1927 *Oil!* has become the singular touchstone, and in some senses a cultural lodestar, for imagining the American oil encounter dramatizing as it does the prodigious birth of the industry in Titusville, Pennsylvania.

Despite the success of Burmah Shell, and the broader network of the Indian Ocean—the setting for Ghosh's first novel, *The Circle of Reason* (1986)—the myth of American exceptionalism as it obtains within the petroleum industry (and surely elsewhere) is bolstered by such limited formulations of petrofiction. This is the thrust of Stephanie LeMenager's 2014 *Oil in the American Imaginary*, in which she documents an American petro-imaginary that also expresses itself in such films as *Giant*. The 1956 film, which dramatizes a shift in labor regimes of the sort documented by Timothy Mitchell—that is, the erasure of human labor through the slippery technics of petrol—also testifies to the persistent forms of mythmaking so central to the American oil imaginary.[18] However much oil figures the present absence of contemporary networks of global capital, "the leading industrialized countries are [in fact] oil states. Without the energy they derive from oil their current forms of political and economic life would not exist."[19]

Notably, *Giant*, based on the 1952 novel by Edna Ferber, was released just two years prior to the export of Nigerian oil by Dutch Royal Shell in 1958. The case of Nigeria provides one of the clearest examples of an American petro-imaginary that foregrounds sites of consumption while sublimating such imperial elsewheres as the Ogoni oil fields. Nigerian novelist Chris Abani's 2004 "petro-picaresque" novel *GraceLand* thus takes on displacement and precarity over and against explicit ruminations about oil, but the evidence of the nation's dependency on "petro-naira" is everywhere.[20] As Matthew Gandy explains, the country's oil boom led to the economic conditions that are foregrounded in the novel—that is, the "dilapidated structures [that] now encircle much of the inner core of the city, casting their shadows across the shacks and stalls that have colonized every available space."[21] So, too, the "onset of global recession in 1981 and the collapse in oil prices" spurred on the sorts of initiatives that Abani's fictional "King of the Beggars" critiques in his indictment of "dose World Bank . . . tiefs."[22]

GraceLand has more often been read as a critique of the IMF-sponsored structural adjustment programs that decimated Nigeria's infrastructure. Read this way, it joins such similar projects as Fidelis Odun Balogun's 1995 short-story collection *Adjusted Lives: Stories of Structural Adjustments*. But to extricate the role of the oil industry from the broader framework of structural adjustment and the tactics of "dose World Bank . . . tiefs" is to deny the material legacy of petroleum in the Niger Delta. Thus, alongside works such as Ken Saro-Wiwa's *Forest of Flowers* (1986), in which his haunting "Night Ride" was first published, or Helon Habila's 2011 *Oil on Water*—also set in an environmentally devasted Niger Delta—we place Abani's *GraceLand.* To read the Abani novel as an oil fiction is to recognize again the limitations of any definition of the genre that focuses merely on the derrick or the pump. Such a reading also forces renewed consideration of figures like Leopold, who might otherwise be laid to rest within an eco-parochial environmental tradition still clinging to the aforementioned Thoreauvian idyll.[23]

In *Oil Fictions,* we are instead interested in reading Leopold with the renewed insights of Abani or in pairing *GraceLand* with a film like *Mad Max: Fury Road* (2014), in which the itinerant Max beholds a sublime desert landscape made so by our disastrous reliance on petroleum. This moment, in many ways, marks the end of Springsteen's "fuel-injected" dream. The elusive "gas town" in the film makes legible the finitude of fossil capital—the imminent dearth so often associated with the Anthropocene—however much conventional end-of-nature narratives foreclose the sort of radical politics that the film's conclusion affords.

The latest *Mad Max* is notable for an ending that turns not on a messianic Max but on a radical collectivism realized at the moment he recedes into the middle distance—an anonymous face amid a newly quenched peasantry. In this sense, the film both stages a productive departure from a petro-imaginary that turns solely on *the* encounter and suggests a philosophical shift aligned with scholars in Anthropocene studies interested in staking out a more robust conception of human and nonhuman agency. The film's geographical ambiguity—the blasted landscape of postapocalyptic Australia made somehow more familiar through a muscle car culture generally associated with the American scene—also allows for a dissolution of the imagined distance between consumption and production as it obtains in the sorts of mythmaking cited earlier.

The present study takes as its point of departure precisely this imagined distance—an otherwise aporic space made palpable by stories like Abani's or like the many fictions taken up here. Oil fictions foreground invisibility, itinerancy, and precarity, materializing invisibility through their scrutiny of the uneven terrain of our global petrosphere. If the discourse around petroculture has thus far been centered on, and to a great extent mediated by, the resource aesthetics of the Global North, the province of *Oil Fictions* is instead a world ecology transformed by petrol.

Oil and World Literature: Turning the Focus on the World

Oil transformed everyday life in the twentieth century. In the twenty-first century, we are finally beginning to realize the degree to which oil has made us moderns who and what we are, shaping our existence close at hand while narrating us into networks of power and commerce far, far away.
—Sheena Wilson, Imre Szeman, and Adam Carlson,
Petrocultures: Oil, Politics, Culture

While the discourse of petroleum has received considerable attention in the Global North, particularly in the United States and Canada, oil's ubiquity is evident in the Global South as well. The Global Energy Statistical Yearbook figures for 2017 show that while the United States and China rank as the highest consumers of oil (demanding 771 metric tons and 530 metric tons, respectively), global oil consumption rates are also significantly high in the Global South, with India, Mexico, Brazil, and Indonesia consuming petroleum on par with postindustrialized European countries, Australia, Russia, and Canada (with a demand between 10,000 and 1,000,000 barrels a day). In 2013, petroleum demand in the developing nations overtook that of the OECD nations. David Sheppard reports that "from around 78 million bpd 10 years ago, global oil demand is now expected to average 89.2 million bpd in 2013, rising by almost 900,000 bpd from last year. . . . Oil demand outside the wealthy nations' club of the Organization of Economic Cooperation and Development (OECD) has jumped by almost 50 percent in the last decade, hitting 44.5 million bpd."[24]

Despite the alarming pollution rates of the Global South, as indexed by studies of the World Health Organization and the *Wall Street Journal*, petroleum use has been mostly normalized as an essential commodity in day-to-day affairs, barring occasional spurts of climate consciousness. In India, for example, while traffic rationing has been introduced in New Delhi to counter alarming rates of air toxicity through vehicular carbon emissions, concerns have been raised regarding the rising fuel price and its unaffordability for the common Indian. In an ominous projection, *Forbes* predicts the dramatic growth of demand for oil in the coming years: "In the US, with 330 million people, the average American consumes 2.7 gallons of oil products a day; in India, with 1260 million people, the average Indian now consumes 0.15 gallons per day. You could do the math on how much oil India could be consuming if they ever consume like us."[25] The rising statistics look more ominous if considered against the high rates of "energy poverty" that still exist in the developing world, where immense disparity affects access to energy, particularly fossil fuels. Greater energy accessibility would invariably call for greater petroleum dependence, unless energy forms are radically revised.

Petroleum's inroad to the developing nations follows as a natural trajectory of the import of the fossil fuel regime to the colonies during the colonial era. The connection between empire and carbon capitalism has often been explored, with similar conclusions by scholars like Paul J. Crutzen and Jeremy Davies, who periodize the Anthropocene as an epoch that began in the late eighteenth century with Britain's exponential growth and the ever-growing role of fossil fuels in the world economy. As Davies asserts, in 1820, half of the world's population came under the British Empire, facilitating a "colonial hinterland" that impelled it toward mechanized and coal-fueled production.[26] Likewise, Christophe Bonneuil and Jean-Baptiste Fressoz propose to use the term "Anglocene" to refer to the immense impact of the imperial policies of Great Britain and the United States in accelerating carbon emissions. As they put it, "The overwhelming share of responsibility for climate change of the two hegemonic powers of the nineteenth (Great Britain) and twentieth (United States) centuries attests to the fundamental link between climate change and projects of world domination."[27] The putative "great acceleration" in climate change, however, occurred over the last half of the twentieth century. Will Steffen and colleagues noted how "nearly three-quarters of the anthropogenically driven rise in CO_2 concentration has occurred since 1950 (from about 310 to 380 ppm), and about half of the total rise (48 ppm) has occurred in just the last 30 years"[28]—a growth that can be attributed to the globalization of neoliberal economic policies. Although Davies sees the petro-regime as a postwar era, or a "golden spike," that "points to a worldwide economic, demographic, and political restructuring, rather than an exclusively European phenomenon," the colonial history of world energy viewed through the import of Western industrial modernity offers a significant critical paradigm of analysis for the contemporary era.[29] Indeed, and as several of the essays in this collection suggest, such facile distinctions between a pre- and postindustrial era ignore a long history of uneven and combined development, which begins well before James Watt's 1784 patent for the steam engine and surely before World War II.

Colonial history and the disparity between the Global North and South inform issues of carbon abstinence as well. The poor countries cannot sustain themselves without cheap, reliable fossil fuels, ensuring a form of carbon dependence that ironically becomes an integral part of the decolonization process of self-reliance. As Tucker Davey aptly points out, "Global warming typically takes a back seat to feeding, housing, and employing these countries' citizens."[30] Thus, the path to more development is paved by increasing the carbon footprints of the poor and middle-income countries, which according to the World Bank already account for just over half of total carbon emissions. The situation seems conducive to a Northern-versus-Southern environmentalism debate about prioritizing carbon abstinence. Awareness of the debilitating environmental effects of the current energy regimes, a critique of energy consumerism, and a quest for alternative energy forms are often read as postmaterialist concerns arising predominantly from the North, where basic amenities are available to a larger population: Ramachandra Guha and Juan Martinez-Alier quote a British journalist who cogently comments that "when everyone turns environmental, prosperity has truly arrived. Greenness is the ultimate luxury of the consumer society."[31] But oil also galvanizes issues of "the empty-belly environmentalism of the South."[32] Northern and Southern environmental concerns have often clashed, the classic case being that of environmental conservation through cordoning off spaces for so-called wildernesses.[33]

Ghosh voices similar concerns about the conflicting priorities of the developed and the developing countries regarding fossil fuel use: "Should we perhaps abandon the quest for Western-style prosperity, so that a greater number will survive? . . . But this would require the abandonment also of the project of modernization that was often implicit in decolonization: it would put a freeze on a system of colonial style inequality."[34] The utter imbalance of power and privilege among geopolitical regions, shaped by empire, and the impetus for decolonization through industrialization creates an impasse of priorities against a looming climate catastrophe. Ghosh proposes a solution that he himself touts as absurd: let the third world develop while the first world sacrifices its energy use until an equitable global balance is reached. Petroleum's relationship with the developing nations has often been touted as "a pathway to a better quality of life"[35] and an integral part of the developmental plan that can be modified for more effective use but cannot be replaced.

What then is the postcolonial take on oil? Does the need for development outweigh the threat of imminent climate disaster? Is the call for alternative energy sources—the chief concern for projects such as Szeman's *After Oil*—mostly a Northern initiative? The dilemma is perhaps not so daunting. As we wrote this introduction, emergency was declared in Borneo after a fatal oil spill, and another massive oil spill has killed thousands of animals and forest land in Colombia; these nations are just five hours apart from each other, asserting petroleum's immense potential for instant global damage. The dilemma of whether interrogating oil is a global prerogative or if poorer nations must first tackle their underdevelopment through cheap available resources is a case of false binaries. The petroleum industry has been particularly successful in naturalizing what Graeme Macdonald calls its "invisible infrastructural element," which sustains the "interests of an oil-reliant capitalism seeking to extend and perpetuate supply while downplaying the ongoing, exploitative shame of extraction and land dispossession and the inevitable endpoints of burning."[36]

The discourse of petroleum (as well as coal) as an inevitable resource for development uncritically embraces the capitalist model of progress, which is based on an extractive economy that abstracts both the sources and effects of extraction to the point of invisibility. Mimicking its material composition, oil operates in a global network of metaphoric liquidity of consumption and extraction, forming a truly transnational web of debilitating structures of domination. Subsequently, given their history of colonial-era exploitation and economic vulnerability in the postcolonial period, postcolonial nations are more susceptible to the detrimental effects of environmental degradation, including those caused by unregulated petroleum use.[37] While D. E. Alexander and R. W. Fairbridge assert the connection between poverty and environmental risks through particular modes of "environmental racism" facilitated by the physical proximity of the poor to hazardous sites like "the flash-floodable canyons of the Andean cities such as Cusco, the steep, unstable tropical hillsides of Caracas, Rio De Janeiro and Ponce, the crumbling and floodable river banks of [the] Ganges and Brahmaputra in Bangladesh . . . , and the sites next to dangerous chemical plants in Mexico City and Bhopal,"[38] the ubiquity of carbon toxicity and the resultant climate change refuse to be mapped against specific, geographically markable sites of disaster.

Climate change is more inclusive, more amorphous, and more global, yet it is more menacing to the world's poor. Lael Brainard and colleagues cogently put it this way: "As the mean temperature of the Earth rises, the impact of climate change on sources of water and food, and on health and living standards, will be greater in those regions that are already struggling," resulting in a "humanitarian disaster on top of the environmental one."[39] Ghosh expresses similar concerns about the imminent climate catastrophe, asserting that "it is not the middle classes and the political elites that will bear the brunt of the suffering but rather the poor and the disempowered."[40] Furthermore, and despite its debilitating magnitude, climate change often plays out as what Rob Nixon calls "slow violence": "a violence that occurs gradually and out of sight, a violence of delayed destruction that is dispersed across time and space, an attritional violence that is typically not viewed as violence at all."[41] For the poor and marginalized, such forms of slow violence—the aforementioned absent presence of oil in its temporally disorienting effects—are particularly devastating.

Although the environmental damage from petroleum has had its share of visibility in haunting media images, such as those of the Gulf War oil spills and the contaminated marine life visually blackened by the disaster, the long-term effects of petroleum use affecting climate change are often "incremental and accretive." The slow violence of petrotoxicity, carbon emissions, oil spills, environmental pollution, and global warming that tilts the developing world toward a greater threat of precarity thus calls for renewed critical attention to the postcolonial engagement with oil in the *Thermocene epoch*—another neologism coined by Bonneuil and Fressoz to describe the political history of carbon dioxide.[42] The discourse on oil has the potential to bridge the classic debate between Northern and Southern environmentalisms, resulting in more collaborative models of environmental issues and resistances.

Apart from the threat of climate disasters, oil awareness in the postcolonial context responds to an exploitative system of extraction facilitated by the systemic violence of imperial forces compounded under state- and corporate-sanctioned neoliberal regimes. In their conclusion to *Varieties of Environmentalism*, a foundational book on postcolonial environmentalism, Ramachandra Guha and Juan Martinez-Alier outline the polemic of the "empty belly environmentalism of the South":

> The environmentalism of the poor, we argue, originates in social conflicts over access to and control over natural resources: conflicts between peasants and industry over forest produce, for example, or between rural and urban populations over water and energy. Many social conflicts often have an ecological content, with the poor trying to retain under their control the natural resources threatened by state takeover or by the advance of the generalised market system.[43]

Conflict over the control of natural resources becomes a cogent trope particularly in the context of the petroleum industry. The volatile power dynamics of control over the oil wealth even in areas of abundance has long been marked as a resource curse of the petroleum-producing states that are scarred by decades of civil war, a lack of democratic systems of governance, and extremely disproportionate wealth. As Michael Ross explicates, since 1980, the oil states around the world—scattered in the Middle East, Africa, Latin America, and Asia—"are no wealthier, or more democratic and peaceful than they were three decades ago."[44] For Ross, these maladies were caused only by oil, not by any other natural resource: "The resource curse is overwhelmingly an oil curse."[45] Timothy Mitchell also argues for the inextricable relationship between "large oil earnings and the difficulty of mounting claims to be more democratic and egalitarian."[46]

Yet, for Mitchell, the predominant rhetoric of an "oil curse" too often focuses on the incapacity of governments to equitably distribute revenues rather than assessing the processes of extraction, processing, shipment, and consumption that make oil into "forms of affluence and power."[47] If the nineteenth-century transition to coal energy, as Andreas Malm argues, was spurred by its superior capitalist potential to exploit and subordinate labor, petroleum involves amorphous networks of transnational capital that feed on rhizomatic structures of extraction involving multinational companies, national governments, and local populations. Graeme Macdonald cogently observes: "As soon as oil is struck, its site is internationalized by virtue of the multinational capital and expertise required (often American) to set up the extraction infrastructure and the labor force, and to enable its immediate plunge into the world market."[48] It is for this reason that Jason W. Moore calls oil "just a goopy substance which becomes the mighty fossil fuel only through modernity's crystallization of human/extra-human nature."[49] Subsequently, anti-oil resistances have erupted globally at the grassroots level by the people most intimately affected by the onslaught of oil in their environments and lives, such as the Standing Rock community of South Dakota over the Dakota Access pipeline, Ken Saro-Wiwa and the Ogoni people in the Niger Delta, and environmental martyrs across the world who are fighting regimes of privatization and enclosure that have decimated local populations. The mantle of resistance to pipelines and megadams is increasingly being worn by activists like Berta Cácares, the Honduran recipient of the 2015 Goldman Environmental Prize, who was murdered for her efforts to thwart the devastation of Native lands.

How has postcolonial literature responded to issues of oil, petro-imperialism, and consequent environmental devastation? Is there a distinct postcolonial genre for oil literature? Graeme Macdonald argues for suspending geographical boundaries when looking for oil fictions, since oil as a resource has an impact both within and beyond the nation-state. Given its global presence, oil is conceived as a theme, as Macdonald asserts, in categories beyond "national literatures."[50] Oil is marked by transnational circuits of literary production in which both texts and literary genres get transported in borderless forms like the oil itself. Macdonald thus looks to world literature to locate the oil novels, which he argues have been published in accelerating numbers both before and after Ghosh's intervention. Macdonald thus points out the omnipresence of oil globally and comments, "Given that oil and its constituents are so ubiquitous in the material and organization of modern life, is not every modern novel to some extent an oil novel?"[51] For Macdonald, the global oil novel, which is often the American oil novel set on foreign shores, contains certain thematic particularities such as volatile "labor relations and ethnic tensions, war and violence, ecological despoliation, and political corruption. Storage and 'peak' anxiety over levels of reserves and remainders shapes events and chronological structure."[52]

Along with themes of conflict, petrofictions also contain emphatic assertions of environmental justice: "Portentous plotlines are common, with the striking of oil and the coming of the oil company men often represented in narratives driven by proleptic inevitability or by a sudden acceleration in events. In such traceable forms, the oil text anticipates the utterly changed world that petromodernity provides."[53] Macdonald draws on an impressive array of world literature focusing on the conflicts of local populations with American oil imperialism—Upton Sinclair's *Oil!*, Ralph de Boissière's *Crown Jewel*, Abdelrahman Munif's *Cities of Salt*, George Mackay Brown's *Greenvoe*, Carlos Fuentes's *The Hydra Head*, Patrick Chamoiseau's *Texaco*, and Nawal

El-Saadawi's *Love in the Kingdom of Oil*. The recurring image of the "sinister corporate interloper," who is often a representative of Big Oil and usually an American, bringing in petromodernity to a resource-rich locale and altering local lives irreversibly, also embodies the combined and unequal development of the oil economy over the world's disparate geopolitical locales. For Macdonald, however, tracing oil in the cultural imaginary of the world doesn't require looking for the obvious markers of oil in the text. Rather, he argues for the ubiquity of oil in all modern literature:

> All modern writing is premised on both the promise and the hidden costs and benefits of hydrocarbon culture. If this proposition seems unwieldy—preposterous even—it is still worth thinking [about] how oil's sheer predominance within modernity means that it is everywhere in literature yet nowhere refined enough—yet—to be brought to the surface of every text. But it sits there nevertheless—untapped, bubbling under the surface, ready to be extracted by a new generation of oil-aware petrocritics.[54]

Oil Fictions is a collective effort to question and contemplate oil through contemporary cultural forms of world literature. Although we are aware of the characteristic globality of oil and the challenges of reading its cultural representations through paradigms of national literature, we have organized our chapters according to their geopolitical contexts. This is a conscious choice made to further the volume's purpose of focusing more on the local dynamics of oil encounters, contextualized against distinct socio-political-cultural milieus of respective nations and their people. There are, however, several texts that spill over national boundaries, offering visions of the oil experience as a transnational, inextricably linked network of people and places operating on a complex scale of power that cannot be mapped by geographical parameters.

Oil Fictions opens with Amitav Ghosh recounting his personal connection with the phenomenon of the oil encounter. In a timely addendum to his article "Petrofiction," Ghosh observes the continued elusiveness of the oil encounter for contemporary fiction writers and suggests new itineraries for thinking about petrofiction and the impacts of petroculture on the literary and political imaginaries of the Global South. Ashley Dawson's essay on organized labor and the intersecting histories of labor and fossil capital addresses what he calls crucial moments of energy system transitions, offering readings of the novels that illuminate their respective geopolitical moments. In this sense, Ghosh and Dawson provide a frame for thinking about the overwhelming forces of petro-imperialism in an era of neoliberal globalization.

Oil Fictions then moves to the African continent, responding to the historical erasure there of labor generally and extractivist networks specifically in Nigeria and elsewhere. African encounters with oil, especially in Nigeria, inform several of the essays in this volume. Sharae Deckard offers a comparative study of Nawal El Saadawi's *Love in the Kingdom of Oil* and Laura Restrepo's *The Dark Bride*, while Helen Kapstein explores the realm of petrofeminism—an inquiry into women in the roles of consumers, producers, and creators of oil—through Nigerian romance novels. Wendy W. Walters then discusses the realities of petromodernity in postrealist literary forms through the speculative fictions of Nnedi Okorafor. The Niger Delta also features in Henry Obi Ajumeze's essay on his reading of Ben Binebai's *My Life in the Burning Creeks* as a theatrical spectacle of environmental decay.

The next section includes essays on South Asian encounters with oil. Editor Stacey Balkan engages with the historical trajectory of fossil fuel transitions in Burma, as documented in Amitav Ghosh's 2000 novel *The Glass Palace*. Editor Swaralipi Nandi looks at petrofictions set in India, focusing on the precarity of the Indian Gulf migrant through Deepak Unnikrishnan's short story "In Mussafah Grew People." Micheal Angelo Rumore then takes up Amitav Ghosh's *The Circle of Reason* (1986) as a representation of the cosmopolitical Indian Ocean imaginary, reading Ghosh's first novel as an instantiation of "petrofiction."

Oil Fictions then moves on to Iran. Simon Ryle reads the Iranian novelist Reza Negarestani's *Cyclonopedia* through an ecopoetic lens, arguing that Negarestani's poetics chart the interconnected materialities of petromodernity through a hybrid politics of human and nonhuman agencies: a "petropolytics." From Iranian speculations on oil, we move on to Latin America. Scott DeVries draws from a voluminous body of Venezuelan and Mexican literary works to argue for the figure of the oil industry as a unanimous posture among authors of South American Spanish literature.

The last two essays of the volume take up texts that attest to petroleum's transnational hegemony through stories that weave through different locales on the globe and that offer comparative forms of the petro-imaginary. First, Imre Szeman, in "Conjectures on World Energy Literature," suggests a theoretical reorientation so as to foreground the uneven and profoundly unequal system of global energy production and to unsettle conventional categorizations of "world literature" that might eschew consideration of the transnational energy regimes taken up in the essay that follows by Corbin Hiday. Hiday analyzes Joseph O'Neill's 2008 novel *Netherland* and Abdelrahman Munif's 1984 *Cities of Salt* as indices of the production of social space within petromodernity. Noting the representational challenges posed by extractive capitalism, Hiday seeks to move beyond mimetic reflection to instead explore the intersections between fossil capitalism and narrative form.

In the final section of *Oil Fictions,* we turn to testimonies of migrant laborers working in the oil fields of the Persian Gulf and the Permian Basin of West Texas. In this sense the volume comes full circle in reexamining the legibility of labor and, especially, informal labor regimes under petro-imperialism. Here Maya Vinai considers the dearth of attention to the experiences and encounters of South Asian migrants to Gulf countries within the discourse around diasporic Indian fiction. She thus turns to Benyamin (Benny Daniel), whose *Goat Days* (originally written in Malayalam and translated into English) is so far the only Indian novel that re-creates the so-called Gulf experience of migrant laborers. In the final essay of the collection, "Testimonies of the Permian Basin," Kristen Figgins, Rebecca Babcock, and Sheena Stief use a collection of memoirs from the Permian Basin community as a starting point for exploring the bioregional impact of energy production on the boom-bust ecosystem. *Oil Fictions* concludes with an afterword by Szeman, whose work on petroculture, petromodernity, and the possibility of cultivating a world "after oil" has been a beacon for the scholars and activists represented here.

Notes

1. The embargo was a response to US intervention in the 1973 war between Israel and, primarily, Egypt and Syria. See also Timothy Mitchell's discussion of "the linked crises of the U.S. dollar and the nationalization of oil in the Middle East" between 1967 and 1974. *Carbon Democracy: Political Power in the Age of Oil* (New York: Verso, 2011), 170.

2. Rob Nixon, *Slow Violence and the Environmentalism of the Poor* (Cambridge, MA: Harvard University Press, 2011).

3. A. R. Ammons, *Garbage: A Poem* (New York: Norton, 1993).

4. Aldo Leopold, "Game and Wildlife Conversation," in *Game Management* (Madison: University of Wisconsin Press, 1933), 23.

5. In *Bodily Natures: Science, Environment, and the Material Self* (Bloomington: Indiana University Press, 2010), Stacy Alaimo traces the co-constitutive patterns of US environmental movements and the "many forms of violence toward and subordination of Native Americans, African Americans, and Mexican Americans" alluded to here (29). See also William Cronon, *Nature's Metropolis: Chicago*

and the Great West (New York: Norton, 1991). In his 2011 *Slow Violence and the Environmentalism of the Poor*, Rob Nixon casts this problem in a more contemporary light by explaining Larry Summers's "poison-redistribution ethic," espoused in his 1991 World Bank memo, as "assum[ing] a direct link between aesthetically unsightly waste and Africa as an out-of-sight continent" (2).

6. Nixon, *Slow Violence*, 16.

7. Brent Ryan Bellamy and Jeff Diamanti, *Materialism and the Critique of Energy* (n.p.: MCM, 2018), 3.

8. Bellamy and Diamanti, *Materialism*. The term *fossil capitalism* derives from Andreas Malm's *Fossil Capital: The Rise of Steam Power and the Roots of Global Warming* (New York: Verso, 2016). Malm defines the "fossil economy," and thereby "fossil capitalism," as such: "An economy of self-sustaining growth predicated on the growing consumption of fossil fuels, and therefore generating a sustained growth in emissions of carbon dioxide" (11). For discussions of the history of the carbon economy, see also *The Shock of the Anthropocene: The Earth, History and Us* by historians Christophe Bonneuil and Jean-Baptiste Fressoz (New York: Verso, 2016), in which they discuss the "Thermocene," or "political history of carbon dioxide."

9. Patricia Yaeger, "Literature in the Ages of Wood, Tallow, Coal, Whale Oil, Gasoline, Atomic Power, and Other Energy Sources," *PMLA* 126, no. 2 (2011): 305–26.

10. Imre Szeman, "Conjectures on World Energy Literature: Or, What Is Petroculture?," *Journal of Postcolonial Writing* 53, no. 3 (2017): 278.

11. Karina Baptista, "Petrocultures," *Global South Studies: A Collective Publication with the Global South*, 2017, https://globalsouthstudies.as.virginia.edu/key-concepts/petrocultures; Mitchell, *Carbon Democracy*, 167.

12. Szeman, "Conjectures on World Energy Literature," 278.

13. In *Global Warming and the Sweetness of Life: A Tar Sands Tale* (Cambridge, MA: MIT Press, 2018), Matt Hern and Am Johal contend, in relation to the Petrocultures project, that "the tar sands are not just an Albertan affair: they are *everyone's* business" (10). We take the term "world ecology" from Jason Moore's *Capitalism in the Web of Life: Ecology and the Accumulation of Capital* (New York: Verso, 2015).

14. Imre Szeman and Dominic Boyer, *Energy Humanities: An Anthology* (Baltimore: Johns Hopkins University Press, 2017), 9.

15. Ibid.

16. Amitav Ghosh, "Petrofiction: The Oil Encounter and the Novel," *New Republic*, March 2, 1992; Ghosh, *The Great Derangement: Climate Change and the Unthinkable* (Chicago: University of Chicago Press, 2017).

17. Ibid.

18. See also Robert Vitalis's *America's Kingdom: Mythmaking on the Saudi Oil Frontier* (Palo Alto: Stanford University Press, 2006).

19. Ibid., 158.

20. See Stacey Balkan's discussion of the "petro-picaresque" in "Rogues in the Postcolony: Chris Abani's *GraceLand* and the Petro-picaresque," *Global South* 9, no. 2 (2016): 18–37.

21. Matthew Gandy, "Learning from Lagos," *New Left Review* 33 (May/June 2005): 44.

22. Ibid., 45; Chris Abani, *GraceLand* (New York: Farrar, Straus and Giroux, 2004), 280.

23. See also Rob Nixon's "Environmentalism and Postcolonialism," in *Postcolonial Studies and Beyond*, ed. Ania Loomba et al. (Durham: Duke University Press, 2005), 196–210.

24. David Sheppard, "Developing World Oil Demand Surpasses Wealthy Nations," Reuters, June 12, 2013.

25. Jude Clemente, "A Steady Drumbeat for More Global Oil Demand," *Forbes*, May 29, 2017.

26. Jeremy Davies, *The Birth of the Anthropocene* (Oakland: University of California Press, 2016), 97.

27. Christophe Bonneuil and Jean-Baptiste Fressoz, *The Shock of the Anthropocene: The Earth, History, and Us* (New York: Verso, 2016), 117.

28. Will Steffen, Paul J. Crutzen, and John R. McNeill, "The Anthropocene: Are Humans Now Overwhelming the Great Forces of Nature?," *AMnio* 36, no. 8 (2007): 614–21.

29. Davies, *Birth of the Anthropocene*, 97.

30. Tucker Davey, "Developing Countries Can't Afford Climate Change," Future of Life Institute, August 15, 2016, https://futureoflife.org/2016/08/05/developing-countries-cant-afford-climate-change/#:~:text=Developing%20countries%20currently%20cannot%20sustain,relying%20heavily%20on%20fossil%20fuels.&text=Yet%20the%20weather%20fluctuations%20and,in%20many%20of%20these%20countries.

31. Ramachandra Guha and Juan Martinez-Alier, *Varieties of Environmentalism: Essays North and South* (London: Earthscan Publications, 1997), xiv.

32. Ibid., xxi.

33. The Northern environmentalist drive for wilderness preservation was severely criticized by Southern environmentalists like Ramachandra Guha, who spoke about the human cost of securing lands for conservation by evicting multitudes of poor forest dwellers and marginal villagers, or "conservation refugees." Mark Dowie, *Conservation Refugees: The Hundred-Year Conflict Between Global Conservation and Native Peoples* (Cambridge, MA: MIT Press, 2009). Amitav Ghosh's 2004 novel *The Hungry Tide* portrays the disastrous consequences of imposing a Northern model of conservation without taking into consideration the local dynamics of the Global South.

34. Amitav Ghosh, "The Great Derangement: Fiction, History, and Politics in the Age of Global Warming" (Berlin Family Lectures, University of Chicago, Chicago, IL, October 7, 2015).

35. Robert Rapier, "Petroleum Demand in Developing Countries," *Energy Trends Insider*, August 2, 2012.

36. Graeme Macdonald, "Containing Oil: The Pipeline in Petroculture," in *Petrocultures: Oil, Politics, Culture*, ed. Sheena Wilson, Imre Szeman, and Adam Carlson (Berkeley: University of California Press, 2017), 39.

37. See Stanley Ngene, Kiran Tota-Maharaj, Paul Eke, and Colin Hills, "Environmental and Economic Impacts of Crude Oil and Natural Gas Production in Developing Countries," *International Journal of Economy, Energy and Environment* 1, no. 3 (2016): 64–73, http://www.sciencepublishinggroup.com/journal/paperinfo.aspx?journalid=349&doi=10.11648/j.ijeee.20160103.13. See also O. Hall, A. Duit, A. Caballero, and L. Caballero, "World Poverty, Environmental Vulnerability and Population at Risk for Natural Hazards," *Journal of Maps* 4, no. 1 (2008).

38. D. E. Alexander and R. W. Fairbridge, eds., *Encyclopedia of Environmental Science* (Dordrecht: Kluwer, 1999), 663.

39. Lael Brainard et al., eds., *Climate Change and Global Poverty: A Billion Lives in the Balance?* (Washington, DC: Brookings Institution Press, 2009), xvi.

40. Ghosh, *Great Derangement*, 148.

41. Nixon, *Slow Violence*, 2.

42. Ibid.

43. Guha and Martinez-Alier, *Varieties of Environmentalism*, xxi.

44. Michael L. Ross, *The Oil Curse: How Petroleum Wealth Shapes the Development of Nations* (Princeton: Princeton University Press, 2011), 2.

45. Ibid.

46. Mitchell, *Carbon Democracy*, 1.

47. Ibid., 2.

48. Graeme Macdonald, "Oil and World Literature," *American Book Review* 33, no. 3 (2012): 7.

49. Jason W. Moore, "The Socio-ecological Crises of Capitalism," in *Capital and Its Discontents: Conversations with Radical Thinkers in a Time of Tumult*, ed. Sasha Lilley (Oakland, CA: PM Press, 2011), 141.

50. Macdonald, "Oil and World Literature," 31.

51. Ibid., 7.

52. Ibid., 31.

53. Ibid.

54. Ibid.

1.

Petrofiction, Revisited

Amitav Ghosh

It was in Iran, back in the days when it was still ruled by the shah, that I became aware of the transformative power of petroleum.

My visits to Iran came about because of my father, who spent a few years there working in the Indian embassy. This was in the early 1970s—I was in school in India then, but I would go to Tehran twice a year to visit my family.

The borders of India and Iran are less than a thousand miles apart, and the two countries are joined by linguistic and cultural ties that go back to antiquity. This made the differences between the countries seem even starker to my youthful eyes. Traveling from New Delhi to Tehran was like crossing a chasm of oil: on the far side lay gleaming highways and fast cars, American-style pizzerias and multichannel television. All of this was new to me, as was the ever-present, always-palpable sense of political oppression that seemed to be the invisible price of petroprosperity.

I went on several trips to the south of the country with my family. We visited many of the towns that sit on the northern shore of the Persian Gulf—ports like Bushehr and Bandar Abbas (which overlooks the Straits of Hormuz). I recall in particular a visit to the town of Abadan, which is on the eastern side of the Shatt al-Arab, the river that divides southern Iran from Iraq. Both banks of the river were packed with oil installations of various kinds—I remember gazing at them in wonderment as they lit up the night sky.

My memories of Iran created a lasting connection with the Middle East. Years later, when a series of fortunate accidents made it possible for me to embark on a doctorate in anthropology at Oxford, it was my early experience of Iran that led me to consider returning to the region. But Iran, having been convulsed by the Islamic Revolution, was then going through a period of turmoil, so I decided to focus on the Arabic-speaking world instead.

In the summer of 1979, after taking an Arabic course in Tunisia, I hitchhiked westward through Algeria and Morocco. In Algeria I had several quasi-surreal encounters with the petro-world. The most memorable of them occurred at the end of a ride that took me to the tiny town of El-Oued, at the edge of the lunar landscape of the Grand Erg Oriental (Great Eastern Sand Sea). I was sitting on my rucksack, wondering where I would spend the night, when a sari-clad woman appeared out of a cloud of dust. It turned out that the local hospital, like many others in the Sahara, employed a good number of Indian doctors and nurses; they gave me a place to stay and regaled me with a grand, home-cooked feast.

My doctoral supervisor at Oxford was an Arabist by the name of Peter Lienhardt. He had served as an advisor to a famously eccentric sheikh of the emirate of Ras al-Khaima in the 1950s; in that capacity he had played a small part in a dispute over the Buraimi Oasis. Peter was a great raconteur, and his stories gave me a sense of personal connection with the phenomenon that I would later describe as the "Oil Encounter."

In the Egyptian village where I eventually went to do my fieldwork I met several men who had worked in the Gulf and regaled me with stories about their encounters with Indians, many of whom were from Kerala, in southern India.

It so happened that my first job was at an institute in Kerala. Rare is the Keralan who does not have a Gulf connection, and I was soon privy to innumerable stories about Indian encounters with Egyptians and other Arabs in various petro-states in the Gulf.

It was in Kerala that I began writing my first novel, *The Circle of Reason*, and it was inevitable, I suppose, that it would draw on these encounters. The novel is set in a fictional oil-rich emirate in the Persian Gulf, and it was the process of writing it that made me aware of how the landscapes of oil resist the accustomed techniques of fiction. All of this weighed powerfully on my mind when I sat down to write the review that became "Petrofiction."

My intention in that essay was to identify the reasons for the literary neglect of the oil encounter. At that time, fiction on this subject was indeed scarce. But today, twenty-five years later, we are in a completely different era, one whose beginnings can be traced back, quite precisely, to September 11, 2001. In the days after the attack it was often said that nothing would be the same again. This has certainly proved to be true of the literary landscape. Over the last fifteen years there has been an enormous outpouring of fiction and nonfiction on subjects that are related in one way or another to the oil encounter: the war in Iraq, the war on terror, September 11, the Muslim experience in America, and so on.

Yet in many ways the oil encounter is still as elusive as ever, because much of the current writing on it is not about an encounter at all but about only one side of it: the Western side. In this sense, today's petrofiction has yet to shrug off the burdens of its past.

April 16, 2018

2.

Energy and Autonomy

Worker Struggles and the Evolution of Energy Systems

Ashley Dawson

Introduction

When the Democratic National Committee (DNC) reversed its two-month-old ban on fossil fuel money in the scorching summer of 2018, it did so with a resolution proposed by Chairman Tom Perez that stated that the party "support[s] fossil fuel workers."[1] The resolution, which reopened the floodgates to donations from "employers' political action committees," was a reaction to what the DNC described as "concerns from labor" that the original resolution "was an attack on workers" at a time when Republicans had been notching up a series of devastating victories meant to deal a final death blow to unions in the United States.[2] Powerful building trades unions such as the International Brotherhood of Electrical Workers, the United Brotherhood of Carpenters and Joiners of America, and the North America's Building Trades Unions (NABTU) are strong supporters of fossil capitalism. For example, when pipeline company TransCanada filed for a permit after the 2016 election to build the Keystone XL pipeline, NABTU issued a press release stating that "on behalf of the tens of thousands of skilled craftworkers poised to earn billions of dollars constructing the Keystone XL pipeline, NABTU are thrilled to support TransCanada's application."[3] But NABTU justified its support for Keystone, which had been the focus of an intense protest campaign by the environmental movement during the Obama administration, not simply by citing the enormous amount of money to be made by fossil fuel workers. The organization also stated that "[they were] delighted that the men and women who make their livelihoods in the construction industry will no longer suffer the indignity of having their chosen careers demeaned as nothing more than 'temporary jobs' by out-of-touch politicians."[4] The press release crackles with class anger, directed in this case against environmentally minded politicians who pointed out that only thirty-five permanent jobs would be generated by the Keystone project, whereas the damage caused by the pipeline would be permanent—at least on any human time scale. The NABTU press release makes it clear that class warfare and climate change are inseparably linked. Given the erosion of wages during the bipartisan neoliberal consensus of the last four decades, it was all too easy for a populist demagogue such as Trump to enroll segments of labor in the cause of climate change denial.

Many sectors of the labor movement are at the leading edge of the fight for climate justice, having taken up positions that are diametrically opposed to those of the building trades. My own union, the Professional Staff Congress, which represents faculty and workers at the City Union of New York, passed resolutions supporting immediate divestment from fossil fuels at the city and state levels in 2012.[5] But the building trades unions wield power disproportionate to their numbers, as their ability to sway the DNC suggests. Their position in support of Trump's planet-destroying policies is grounded in pragmatic concerns: welders and other pipeline workers earn $22 per hour on average, while the median hourly wage for solar installers is just $16.[6] Only 4.4 percent of workers in coal, oil, and gas extraction are union members, but the renewable energy sector is almost entirely nonunionized. This has led to deep skepticism among labor unions around the world about the notion of a just transition. Although the idea of *just transition* emerged within the labor movement—when the Oil, Chemical, and Atomic Workers International Union fought to establish a "Superfund for Workers" after a toxic chemical facility in New Jersey was shut down in the mid-1980s—the term has gained traction with the climate justice movement in recent years, where it is used to refer to the need for a broad economic and social transition to a sustainable, zero-carbon world in the next decade or two.[7] Although the social revolution the planet needs should also be a boon for working people the world over, all too often the immediate concerns of workers have not been adequately considered by the environmental movement. And, conversely, segments of the union movement have been ensnared in dominant ideologies of "green growth" and, worse still, climate change denial. Despite the significant erosion of the power of worker organizations across North America and Europe over the last several decades, they remain a key site of counterpower to the capitalist class. It is therefore imperative that notions of green growth as a panacea for the working class be roundly debunked and that calls for a worker-focused transition be built on radical programs that will bring the kind of sweeping socioeconomic transformation that climate science has demonstrated we need in order to avert planetary ecocide.

As we develop this strategy and the policies through which it will be implemented, it is important to recall the pivotal role played by workers during past moments of transition from one energy regime to another. Literary works are particularly interesting documents of the concerns that arise at such historical inflection points. Although their depiction of energy workers is often indirect and heavily slanted, literary representations nonetheless offer an important index of the power held by workers and their organizations during these moments of transition. Novels in particular can be said to dramatize the political unconscious of fossil capitalism during times of crisis and transition. In what follows I will discuss three such pivotal moments of energy system shift, looking at a literary work that in each instance offers telling insights into the fears and desires of petrocultural worlds in transition. In the first of these moments, during the twilight years of coal's dominance in Britain, H. G. Wells's *The Time Machine* bodied forth uncanny images of a working class subjected to terrifying evolutionary pressures. In the second moment, when Saudi Arabia was being incorporated into petromodernity, Abdelrahman Munif's *Cities of Salt* depicted the nomadic world of the Bedouin subordinated to wage slavery. And in the third moment, during the rise of the extreme extraction technology known as fracking in rural Pennsylvania in the 2000s, the itinerant workers who inhabit company-funded "man camps" are counterposed to the dejected locals who reside among the ruins of successive waves of petromodernity. When seen through one lens, these three historical snapshots may seem to tell a story of the smashing of worker power and the consequent exorcism of fears of revolutionary responses to petroculture, but they could also be seen as begging nagging questions about the need for the wholesale transformation of petrocultural worlds that amass mammoth, unsustainable riches for a few while leaving very little for increasing numbers of people.

Energy Transition and Worker Struggles

Shifts from one energy system to another, energy historian Bruce Podobnik argues, are catalyzed by three key dynamics within capitalist cultures:

commercial competition, geopolitical rivalry, and social conflict.[8] On the most fundamental level, capitalists and the corporations they form are locked in a remorseless struggle with one another. The cutthroat competitiveness of the individual capitalist is not simply a product of his individual psychological constitution—although capitalist culture does reward brashly competitive behavior to the detriment of other social values, such as cooperation—but rather a structural feature of the system as a whole.[9] If the use of particular energy resources can give the capitalist a leg up on his competition, they will be adopted and employed to the hilt (as part of a broader repertoire of competitive tactics and dirty tricks). Adversaries who do not appropriate the new energies and technologies will be put out of business with no remorse but instead with much chest thumping and discussion of entrepreneurial zeal. So inasmuch as steam power permitted more intense exploitation of workers, it ultimately triumphed over water power. Similarly, as we shall see, when petroleum proved more useful than coal in the competitive world of capitalism, a shift in energy regimes took place with remarkable speed.

In addition, since capitalists also organize themselves into a national bourgeoisie, the same competitive dynamics we see between rivalrous firms play out between different nation-states. If a particular energy regime and its connected infrastructural assemblage give a country a competitive edge in either business or military affairs, that energy regime is likely to be quickly adopted and exploited. Nations that do not follow suit are dumped unceremoniously into the dustbin of history, turned into vassal powers, and denigrated ruthlessly as culturally benighted and even degenerate.[10] In the nineteenth century, for example, Western European governments, recognizing the importance of steam power in increasingly industrialized forms of warfare following the Napoleonic Wars, strongly promoted the expansion of coal industries. As the century progressed, the world was knit together into a global energy system based on coal. Yet if coal-based steam power was instrumental in the creation of dense, industrialized, networked cities like London, Paris, and New York, it also facilitated the domination of far-flung colonies, as steamships and railroads, in conjunction with other technologies of the era such as the telegraph and the Gatling gun, led to the conquest of vast territories around the world.[11] Coal power thus helped produce a "great divergence" within the world system.[12] At one end of this system were liberal (but imperial) democracies like Britain; at the other end were colonial dominions such as India, China, and the ailing Ottoman Empire. With its technologically backward and cruelly exploitative Southern plantation economy and its isolated reservations of brutally subjugated Indigenous peoples, the United States unified these opposing poles of the world system in one continent-spanning nation, with correspondingly incendiary political implications.

But it is not just the competitive dynamics of the capitalist world system that catalyze energy regime shifts. The third major factor that prompts such transformations is made up of social mobilizations and conflicts. These are at times the result of uprisings by workers within particular energy industries but may also occur within the sprawling infrastructural assemblages that particular forms of energy make possible. Often, these struggles are knit together, as workers in one portion of the energy assemblage are emboldened by and rise up in solidarity with workers in another. In order to defeat these uprisings, private enterprises and capitalist states may seek out a new energy regime that allows more flexible forms of production and distribution.

This is precisely what happened with coal-based steam power. Just as this energy regime was achieving global primacy in the late nineteenth century, coal production in Western Europe and the United States was disrupted by waves of labor militancy that lasted several decades. Conditions in the coal mines were deplorable, with occasional terrible disasters such as cave-ins and dust- and gas-related explosions leading to hundreds of deaths. Wages for mine workers were often below subsistence level, and in the United States one quarter of the workforce in mines was made up of boys.[13] In response to these dangerous and oppressive conditions, miners formed some of the most militant unions in the industrialized world. Despite the violent repression to which these worker struggles were subjected, newly formed miners' unions enjoyed a significant strategic advantage: coal had to be transported up out of the earth and then to urban markets by rail and barge. Striking miners were joined by militants in the connected rail, docking, and shipping industries in a wave of strikes that lasted from the 1880s until the outbreak of World War I. By shutting down the channels through which coal was transported to industrial cities, these workers gained tremendous political leverage, as Timothy Mitchell documents in *Carbon Democracy*.[14] Miner demands centered not just on amelioration of their pay and working conditions but also on far more sweeping democratic reforms in Europe and the United States. Miners were certainly not alone in their militancy: suffragettes, socialists, and anti-imperialists, among others, were all pushing for social transformation in the liberal, imperialist democracies around the beginning of the twentieth century. These struggles were most threatening to the established order when they converged. Indeed, as the revolutionary political organizer Rosa Luxemburg argued in *The Mass Strike* (1905), the economically based struggles of workers could intersect with and strengthen revolutionary political struggles in other sectors of society, since the various forms of the mass strike "run through one another, run side by side, cross one another, flow in and over one another [in] a ceaselessly moving, changing sea of phenomena."[15] The struggles of workers in the coal fields were especially threatening to elites, since they choked off the sources of power that fueled industrial civilization.

Worker power was so threatening to the established order, and their rebellions against the capitalist system were so incomprehensible to its economic architects and intellectual guardians, that workers were represented as fundamentally alien. The illustrations created by middle-class Victorian voyeur Arthur Munby of female coal workers in northern England underline the titillating alterity of energy-sector labor. The women pit brow workers are short, squat, and black-colored in face and limb. They appear to be almost of a different species than the tall, thin, and completely white members of the bourgeoisie with whom Munby frequently juxtaposes them for shocking effect. Anne McClintock has argued that Munby's images draw on social Darwinian tropes, broadly accepted among members of the Victorian middle class, about the racial degeneration of the British working class.[16] That is, Munby's drawings suggest that work in the mines provoked a transformation in members of the working class such that they came to resemble putatively less evolved Africans. By burrowing deep into the entrails of the earth they have also regressed in time, falling back down the evolutionary stepladder to the point that they share little in body and intellect with the nation's bourgeoisie. Munby's images exacerbate powerful anxieties about class difference that are also animated by fears about the purported racial divides observed in an imperial context. Ironically, these tropes of difference, germinated in order to legitimate imperial power, came back to the metropole to haunt the Victorian elite as British coal miners became increasingly militant in their demands not just for higher wages but also for a leveling of the gaping social hierarchies of the age.

H. G. Wells's first novel, *The Time Machine* (1895), crystallizes many of these anxieties about the direction of history and the character of the working class. Wells had won a scholarship to study biology in London under T. H. Huxley, a well-known protégé of Darwin, during which time Wells began to write works of speculative fiction that explored then-popular ideas of degeneration. *The Time Machine* opens with a framing narrative in which an unnamed narrator describes a visit to the home of a well-known amateur scientist, who explains to the narrator and the rest of the skeptical company he has assembled that he has discovered how to move about in time, the fourth dimension, just as others are able to move about in space. When the skeptical guests assemble again a week later, the scientist stumbles into the room in a disheveled state and narrates the tale of his travel in time. The scientist, catapulted eight hundred thousand years into the future by his machine, finds himself in a civilization populated by the gentle, childlike Eloi. These people, he speculates, are the descendants of the communist society whose doctrines haunted bourgeois European cultures at the time that Wells was writing. The scientist guesses that, free of any competition to survive as a result of the triumph of communism's egalitarian ethos, humanity must have degenerated into a condition of puerile fragility.

As he explores this civilization of the future, however, the time traveler realizes that his initial hypothesis was flawed. The feckless simplicity of the Eloi, he learns, is sustained by the labor of the Morlocks, hideous creatures who live and toil in subterranean warrens beneath the brightly sunlit world of the Eloi. Wells's time traveler here extrapolates from the increasing class conflict that characterized the beginning of the twentieth century to represent a world in which the dominance of the bourgeoisie has become so absolute that two separate races have evolved. The Morlocks, who manufacture the products on which the Eloi depend, might be said to represent the proletariat in general. But their subterranean world of giant machines, which seems in some way to provide power for the carefree Eloi, suggests that they bear more than a passing resemblance to the coal workers whom Munby depicted as degenerate brutes. Wells's time traveler articulates precisely such fears of degeneration after he climbs down into the world of the Morlocks. He barely escapes alive but sees enough to realize that the tables have turned: the Morlocks are not the oppressed class, haplessly exploited by the Eloi. Instead, the Morlocks are a race of cannibals who venture out at night to gorge on the Eloi, whom they keep alive like placid fatted calves. The instinctive sympathy the time traveler felt for the Eloi, exemplified most explicitly in his taboo-defying attraction to a childlike female named Weena, is heightened as he realizes that he too has now become the Morlocks' prey. Although the time traveler manages to escape this demise by stealing his time machine back from the Morlocks, his fate grows only bleaker as he lurches even further forward in time. Voyaging thirty million years into the future, the time traveler witnesses the grim scenario about which Lord Kelvin speculated earlier in the nineteenth century: the heat death of the sun. The specter of energy-worker rebellion thus comes to overlap with fin de siècle fears about the unwinding of the universe's primal energies.

The American Century and the Control of Oil (Workers)

Just as workers were gaining significant power through mass action in the coal-based energy system, however, a new petroleum-based energy regime was being born. Commercial drilling for oil had begun in various parts of the world in the mid-nineteenth century. Edwin Drake's 1859 well near Titusville, Pennsylvania, is often considered the first modern well, but the oil industry was a global enterprise from the start, with drilling going on in Britain, Poland, and Azerbaijan, among other places. This oil was mostly refined into kerosene to be used for lighting in homes and streets. Oil may have been lucrative, but it did not initially appear to be the base for an energy regime that could compete with coal. Indeed, the 1876 Centennial Exhibition in Philadelphia devoted little attention to petroleum-based technologies. But oil quickly assumed pivotal geopolitical significance as an interimperial rivalry between Britain and Germany emerged in the late nineteenth century. As a result of petroleum's high energy content, oil-powered ships could travel twice as far as coal-powered steamships, and because of the ease of transporting oil, ships using it could refuel at sea, giving an oil-powered navy decisive strategic advantages.[17] After the passage of the German Naval Law of 1900, which called for the construction of a new oil-powered fleet of battleships and cruisers, Britain found its global naval hegemony imperiled. Under the leadership of Winston Churchill, the British navy responded to the German buildup by beginning an aggressive campaign to switch its fleet to oil. This in turn necessitated a fresh round of imperial expansion, since Britain had abundant domestic coal reserves but no petroleum. In the summer of 1914, just as World War I was beginning, Britain acquired a 51 percent stake in the Anglo-Persian Oil Company, which agreed to provide the British navy with a twenty-year supply of oil on favorable terms. Thus began the West's long history of oil-fueled imperial meddling and warfare in the Middle East.

The unique material characteristics of oil—the ease with which it can be extracted from the ground and transported by pipelines and ships—was key not just in interimperial competition but also in routing the militant coal workers of Europe and in defeating demands for self-determination in oil-rich nations such as Iran and Iraq. In contrast with coal, oil requires a relatively small labor force; all you need, in the iconic image of the wildcatting oil prospector, is a ramshackle drilling rig and a bit of luck, and you can hit a gusher that makes you rich beyond your wildest dreams. Oil also flows easily through multiple channels, making it far more difficult for workers to choke off supplies as coal miners and their allies were able to do by occupying strategic shipment points in the coal infrastructure. The invention of infrastructures such as the oil pipeline, bulk tanker, and large storage tanks meant that oil could be transported almost without manual labor, a stark difference from the coal assemblage. When workers tried to choke off supplies of oil, companies could simply shift the conduits through which oil was moved.[18]

Oil played an important role in defeating popular struggles after 1945. When coal miners in France's Communist Party–led union movement went on strike shortly after World War II in response to rapid inflation, the newly dominant United States organized an aid package known as the Marshall Plan, a key ingredient of which was the conversion of Europe's energy systems from coal to oil.[19] This shift in energy regimes decisively weakened the power of the organized working class in Europe. In addition, since Europe lacked significant quantities of oil, arrangements were put in place for American companies to ship oil in from the Middle East. Most of these oil supplies came from Saudi Arabia, where US oil companies and the American government were intent on propping up the unstable regime of Ibn Saud. Since payments for Saudi oil were made in dollars provided through the Marshall Plan, the recovery package also conveniently established the dollar as the basis of the global financial system. On the back of these US subsidies, oil's share of Western Europe's energy consumption went from one-tenth in 1948 to almost one-third by 1960.[20] The rise of the oil-based energy regime was thus driven to a significant extent by elite efforts in the United States and Europe to quash the power of militant workers in one of the economy's most strategic sectors, but it was also motivated by and inextricably linked to the rise of a new geopolitical hegemony. The shift to a new global energy regime coincided with and helped reinforce the new American Century.

Contrary to the concerns about an oversupply of oil that have circulated in recent years in tandem with the notion of peak oil, the foundational strategy of the American Century was a determination to keep supplies of oil limited and prices cheap and predictable. This entailed a struggle to control and limit production of oil both domestically and on the international stage. In the late nineteenth century, there was only one significant oil-exporting region in the world: northeastern Pennsylvania. By controlling the refineries through which this oil flowed, one company, Standard Oil, came to dominate the global commerce in oil. But that quickly changed, particularly as European oil companies began discovering oil in the Middle East. Early in the twentieth century, oil companies such as Royal Dutch, Shell, Deutsche Bank, and Burmah Oil prodded local rulers into granting them concessions to oil discovered in Egypt, Persia (known as Iran after 1935), the Ottoman provinces of Baghdad and Mosul (which became part of the state of Iraq in 1920), and other parts of the Middle East. When Standard Oil found that it could not destroy these competitors, it instead came to terms with them, creating a Western cartel that worked to restrict the worldwide production and marketing of oil and to guarantee that nascent Middle Eastern nations would not upset the cartel's tightly controlled arrangements through independent oil production.[21] The initial aim of oil prospecting was thus, counterintuitively, not to produce more oil but rather to keep fresh supplies off the market. Companies that discovered oil in places such as Mosul would simply not pump it up from underground.

When anticolonial struggles spread through the Middle East in the early to mid-twentieth century, the United States played a key role as the world's swing producer, capable of increasing production when governments tried to gain control of their oil supplies through nationalization. This allowed such upstart regimes to be isolated and their freshly nationalized oil supplies to be kept off international markets. For instance, after the Iranian government under Mohammad Mossadegh took over the Anglo-Iranian Oil Company (formerly the Anglo-Persian Oil Company) in 1952, the United States increased production in order to enable an effective international boycott of Iranian oil.[22] When these measures did not succeed in reversing the nationalization of Iranian oil, the CIA orchestrated a coup that overthrew the Mossadegh administration, funneling millions of dollars into the country to buy off Mossadegh's supporters and to bankroll street protests against his democratically elected government.[23] Under US oversight, Mohammad Reza Pahlavi was installed as shah, ruling Iran with increasingly despotic means and serving as America's primary military surrogate in the Gulf region until the revolution of 1979. In some cases, the threat of violence against the efforts of anticolonial nationalists to gain control of the energy commons was even more naked: in 1945, for instance, FDR struck a deal with Ibn Saud that committed the United States to intervene militarily to defend the Saudi regime from internal and external aggressors.[24] In sum, powers such as the United States and the oil companies based therein deployed a panoply of measures, from quiet diplomacy to crushing military force, to govern the global energy commons.

Abdelrahman Munif's Cities of Salt fiction series offers an epic account of this sordid history of petro-imperialism and the rise of tyrannical puppet regimes in the Middle East, one that scathingly debunks the anodyne propaganda of American oil companies concerning their role in bringing development to the region. The titular first volume of the quintet chronicles first the destruction of life in a Bedouin oasis following the discovery of oil by prospectors from the United States and then the transformation of a coastal town into a refinery complex staffed by American oilmen and the displaced and hyperexploited Bedouin who have been recruited to work there. The Americans remake the town of Harran in the image of a US Jim Crow city, with a luxurious, gated suburban enclave complete with swimming pool, cinema, and air-conditioned bungalows for the white workers set against a bleak compound consisting of intolerably hot metal shacks for the Arab workers. But the physical privations experienced by the Bedouin men dragooned into the wage economy pale in comparison with the cultural and emotional slights heaped on them by their new American overlords. As Robert Vitalis has documented, the American oilmen brought with them attitudes and comportments forged not simply by the racist mores of the US plantation economy but also by the longer history of settler colonialism in the Americas.[25] While documenting Bedouin culture with ethnographic precision, the oilmen also mock and belittle the disorientation of the Arab workers as they struggle to master new regimes of labor in an unfamiliar cultural milieu. The workers are forced to wear Western clothes entirely inappropriate for the local climate; they are subjected to demeaning identity checks and pumped for useful cultural information; and they are roundly abused by a new comprador class of supercilious Arab translators, merchants, and local rulers. When the workers complain about losses such as a young Bedouin man who tragically drowns, the Americans refuse to recognize, let alone compensate, them. The humiliations accumulate gradually, until the simultaneous death by beating of the town's traditional healer and the company firing of a large group of workers provoke a protest movement. The Arab workers of Harran go on strike, demanding the reinstatement of their colleagues and an honest investigation of the healer's death.[26] Strengthened by rumors of the reappearance of the desert rebel Miteb al-Hathal, patriarch of the uprooted oasis, they overwhelm a detachment of trigger-happy soldiers, free a group of detained workers, and force the technology-addled local emir to flee.

In his seminal essay on "Petrofiction," Amitav Ghosh argues that Munif's depiction of the successful strike is the major weakness in his otherwise dead-on account of the myriad indignities visited on the Bedouin populations of Saudi Arabia following the discovery of oil.[27] As Rob Nixon astutely points out, Munif's novel is indebted to novels of collective struggle such as Emile Zola's *Germinal* and Ousmane Sembene's *God's Bits of Wood*, works that depart from the tradition of the bourgeois novel by featuring multiple protagonists whose oppressed state ultimately culminates in explosive social rebellion.[28] But instead of celebrating this formal departure and political commitment, Ghosh argues that Munif's residence in the former Yugoslavia during the Tito era influenced him to conclude his novel with a saccharine working-class victory. Ghosh is correct inasmuch as the Dhahran strike of 1953, which clearly serves as an inspiration for Munif's novel, was quashed. When Aramco's Saudi workers struck for higher wages and better housing in Dhahran, which by that time was also the site of an American air force base, the monarchy sent in troops and forced the workers to return to work.[29]

But the Dhahran strike is not the only historical antecedent for Munif's novel, which was published in Arabic in 1984: the Iranian Revolution had convulsed the Middle East during the years just before the publication of Munif's work. To the extent that Ayatollah Ruhollah Khomeini appropriated the populist anticolonial rhetoric of Iranian intellectuals such as the Fanon-influenced Ali Shariati, the Iranian Revolution and the regime established in its wake constituted a profound menace to US client regimes such as the Saudi monarchy.[30] The threat was underlined when, not six months after the Iranian Revolution's successful ousting of the shah, a group of 250 heavily armed militants seized the Grand Mosque in Mecca during the annual hajj, taking thousands of pilgrims hostage.[31] The attackers were Sunnis and had no direct links to Shi'ite Iran, but, clearly inspired by the Iranian Revolution, they denounced the Saud family's lack of Islamic rigor; extravagant Lamborghini-driving, polo-playing lifestyle; and subservience to America. The monarchy sent in the military, leading to weeks of heavy combat and carnage that culminated in the capture and beheading of sixty-three rebels. This event, which took place during the Iranian hostage crisis, was largely ignored by the international media but inflamed Muslim rage against the United States, whom Khomeini blamed for the events. Following the attack, King Khalid of Saudi Arabia implemented a stricter enforcement of religious law and gave religious conservatives more power in the kingdom in an effort to neutralize the messianic populism evident during the seizure of the Grand Mosque. These measures were insufficient, however, to quell these populist currents, which led ultimately to the formation of al-Qaeda and subsequent epigones.

The uprising that concludes *Cities of Salt* is not animated by the kind of secular workerist dogma that Ghosh rather derisively attributes to Munif. Rather, the anger of the workers is stoked by the anti-imperial denunciations of a religious character named Ibn Naffeh. Initially introduced as a risible crank, someone out of step not simply with the new town being created by the Americans but also with the canny strategies of merchants such as Ibn Rashed, Ibn Naffeh becomes a pivotal character in the novel's depiction of the popular uprising in Harran. He blames the murder of the town healer on the Americans, and as the people of the town gather for a protest march, he delivers a speech after Friday prayers at the mosque in which he reminds them that it is the duty of the Arab people to resist oppression. It is the shooting of Ibn Naffeh during the march that leads the workers and their allies among the townspeople to storm the oil company compound and free the captive workers. Ibn Naffeh, who survives the shooting, is given the novel's final word, when, following the ignominious flight of the emir from Harran, he pronounces that the Americans are the source of the people's problems. The novel's concluding message is thus not so much one of traditional proletarian solidarity and struggle over working conditions as it is one of zealous anti-imperialism cloaked in religious and populist rhetoric. Munif's novel should therefore be seen not as a distorted account of a decades-old strike but rather as a telling rendition of the rise of insurrectionary Islam on the back of frustrated desires for democracy and simmering anger at the collusion of local petro-despots with US empire. If the American Century was predicated on control of global oil supplies and the client regimes that bring the devil's excrement out of the earth and to market, it would prove to be a dominion just as unstable and unsustainable as the cities of salt that provide the title of Munif's great fiction series.

Prisoners of the American Dream

In one of the most influential theorizations of working-class power during the high era of petroculture, Antonio Negri argued that labor is characterized by an ineradicable *autonomy* from capital.[32] For Negri, labor is its own self-generating power or force—called *potenza* in Italian—that emerges from the bodies and minds of workers. It is consequently autonomous from capital and cannot be fully subordinated to the domineering designs of the bourgeoisie. No matter how powerful and apparently victorious the capitalist, workers will always retain residual traces of autonomous subjectivity that can reject his authority. Negri made these arguments while participating in the uprisings of the *anni di piombo*, or "years of lead," in Italy during the 1970s, when worker struggles grew increasingly intense on factory floors, spilling out into the streets and joining with many broader social movements that challenged the enduring inequalities in capitalist societies.[33] Yet just as organizations such as Potere Operiao in Italy and the Dodge Revolutionary Union Movement (DRUM) in Detroit were fighting for greater worker control on the factory floor and over the social machine more broadly, capital was experimenting with novel ways to crush worker autonomy. Fossil capital was particularly successful in its war against workers. In the Middle East, as Ghosh notes in his critique of Munif, this war took the form of an amplification of the Jim Crow tactics of racial stratification documented in *Cities of Salt*. To keep the number of Arab workers in the oil industry low, the oil sheikhdoms imported large numbers of migrant workers from the poorer countries of Asia. As Ghosh memorably comments, this created "a class that is all the more amenable to control for being perpetually under threat of deportation . . . , a class of helots, with virtually no rights at all."[34] Through the creation of a new international division of labor, in other words, capital undermined—at least temporarily—any capacity of workers to act autonomously in a particular national context. If capital (and oil) could flow freely across political boundaries, the movement of workers was tightly controlled in order to retain the most easily exploited reserve army of labor.

While similar strategies were being experimented with in Western Europe and North America, where so-called guest-worker programs helped provide a reserve of insecure labor, fossil capital also amplified its use of another weapon of class war during this period: automation. In response to strikes by the Oil Workers International Union in refineries along the US Gulf Coast, oil companies introduced machinery that made the production process increasingly autonomous of human beings.[35] For example, as Matt Huber documents, by the late 1950s computer technologies were being used in refineries to monitor and control refining processes. As such automation took hold, strikes not only became less effective but also allowed fossil capital to explore possibilities for making human labor redundant. As an article in the *Oil and Gas Journal* crowed, "One of the ironies of the strikes has been the discovery by the companies that they could run the refineries with still fewer people without sacrificing safety or efficiency."[36] The effectiveness of this strategy and the helplessness of workers to combat automation was made evident during a strike by the Oil, Chemical, and Atomic Workers International Union at a Shell refinery in 1963. Shell was able to operate the refinery at near full capacity throughout the year-long strike, and the union was forced to agree to a very modest wage increase in exchange for the firing of nearly a quarter of the workforce.[37] Fossil capital had hence made significant strides toward making itself autonomous of labor, rather than vice versa, and had turned the worker into human surplus in the process.[38] It is not surprising that this process of automation should occur first and foremost in the domain of fossil capital. After all, capital has historically been dependent on fossil fuels in order to chart what George Caffentzis calls "technological paths to repression."[39] In other words, in order to break workers' power over the production of commodities, capital depends on fossil fuels to provide counterenergy capable of powering the machines that substitute for human labor. Fossil energy, then, is a decisive factor in capital's efforts to win the class struggle.

Jennifer Haigh's novel *Heat and Light* (2016) delivers a bleak but nonetheless engrossing and often amusing exploration of the tragic fragmentation of US society in the face of the most recent avatar of fossil capitalism: the fracking industry. Like Munif's *Cities of Salt*, Haigh's work is a choral novel, with myriad characters who occupy a multitude of different social positions, from victims to beneficiaries to agents of the fracking industry. The novel is set in rural Pennsylvania, where the Texas-based Dark Elephant Energy Company has begun buying up rights to subsurface minerals and drilling wells to get at the oil and gas deposits made newly accessible by hydrofracking technology. Haigh draws witheringly satirical portraits of Texas oil barons like Kip "The Whip" Oliphant, who spouts New Age bromides such as "Now is the time to leverage our first-mover advantage" while he puts his firm ever deeper into debt in order to get at the rich load of carbon sequestered in Pennsylvania's Marcellus Shale.[40] But the novel is most powerful in its depiction of the predatory relationship between the hapless local residents of the town of Bakerton and the emissaries of the fracking industry who arrive to suck their land and their future dry.

A perfect example of such predation is found in the character of Bobby Frame, the scrupulously clean-living Mormon con man sent by Dark Elephant to convince unwitting farmers to sign leases that permit the company to begin drilling on their land. Frame, Haigh writes, doesn't care a whit about the carbon-based boom-and-bust history of Bakerton that has left locals so penniless and thus so vulnerable to his promises of quick and easy riches. Frame's mild manner and apparently artless patter—"Beautiful property you've got here"—conceal ambitions to exploit the land and those who reside on it to the hilt. Haigh also gives us characters such as Herc, the foreman of the Dark Elephant drilling crew. Although Herc refuses to live in the dispiriting man camp set up by the company, he cannot escape the alienating impact of an oilman's rootless life. His marriage in Texas foundering, he takes up with the lonely Pastor Jess in Bakerton but of course neglects to tell her that he's married. His silence and exploitation of Jess serves as an allegory of the broader exploitation of Bakerton's hapless residents by fossil capital. In perhaps the most discouraging thread of *Heat and Light*, Haigh suggests that even antifracking activists such as the geologist Lorne Trexler are interested in local people only to the extent that they can be used to further their own agenda. Although *Heat and Light* offers glimmers of hope—such as the coming to political consciousness of Rena, whose life on an organic dairy farm with her lesbian partner, Mack, is menaced by the arrival of fracking—there is precious little genuine solidarity and no viable resistance to be found in the novel. Indeed, it is the vagaries of the stock market rather than any popular uprising against fracking that ultimately bring Dark Elephant down.[41]

Such pessimism is a telling reflection of the demoralized state of progressive forces in the United States. Labor unions in the United States, for example, have reacted to the dramatic deterioration of their position, which has resulted from a combination of technological changes like automation and a full-frontal political attack waged in the courts and legislatures since the 1970s, by adopting a blinkered defense of workers in the most secure sectors of the economy. They have thereby amplified capital's traditional strategy of segmenting the labor force in order to play workers off one another. As Mike Davis puts it in his blistering critique of the labor movement, "Unions have closed in around the laager of the seniority system, abandoning the unemployed, betraying the trust of working-class communities, and treating young workers as expendable pawns. Such a blinkered, Maginot-like defense of existing employment privileges risks the creation of a reactive anti-solidarity, as the unemployed become strike breakers, or the second-class citizens of the lower wage tiers decertify unions that have failed to represent them."[42] The support of groups such as NABTU for the policies of the Trump administration, including the build-out of planet-destroying pipeline infrastructure, is a sorry example of the extent to which relatively elite workers are willing to put short-term interests before solidarity with other workers who have stood against fossil capitalism. To take this stance is also to reject solidarity with frontline communities in places like Standing Rock in the United States, to ignore the impact of further fossil infrastructure on vulnerable populations around the world, and to become complicit in the unfolding tragedy of planetary ecocide more broadly.

But the capitulation of powerful sectors of the labor movement to fossil capital has not happened in a void. In 2008, inspired by neoliberal Thomas Friedman's use of the term, Barack Obama promised workers in the United States a "Green New Deal."[43] But Obama and the Democrats' tepid market-oriented proposals included items such as the Waxman-Markey cap-and-trade bill, which proposed a gamut of corporate-friendly half measures that would have achieved minimal reductions in emissions. It satisfied no one and went nowhere. Promises to create millions of well-paying green jobs evaporated as fiscal austerity became dogma, and Obama shelved the Green New Deal rhetoric after the midterm elections. It is therefore not surprising that what's left of the labor aristocracy should be leery of promises about green jobs. Today the idea of a Green New Deal is being revived, this time by overt socialists like Alexandria Ocasio-Cortez. Their proposals are far more sweeping and ambitious and include a determination to reach 100 percent renewable energy in a relatively short time frame and a complete rebuilding and climate-proofing of US infrastructure. This time, the Green New Deal, we are promised, will be of the magnitude of the mobilization for World War II and will include a federal green-jobs guarantee and trillions of dollars of spending calculated to transform the infrastructure of every sector of the US economy. Although few are yet talking about this, one way to fund these ambitious proposals would be through the nationalization and gradual, calculated liquidation of Big Oil. We must be clear that it is only by smashing fossil capitalism that we will end the long war against workers and the planet.

Notes

1. Alexander Kaufman, "Democrats' Drama on Fossil Fuel Money Shows a Radical Green Jobs Plan Can Be a Win-Win," HuffPost, August 11, 2018.
2. Kaufman, "Democrats' Drama."
3. NABTU, "North America's Building Trades Unions Welcome TransCanada Refiling for Keystone XL Approval," press release, https://nabtu.org/press_releases/north-americas-building-trades-unions-welcome-transcanada-refiling-keystone-xl-approval/.
4. Ibid.
5. Nancy Romer, "How New York City Won Divestment from Fossil Fuels," *Portside*, March 2, 2018.
6. Kaufman, "Democrats' Drama."
7. On the history of the term *just transition*, see Sean Sweeney and John Treat, "Trade Unions and Just Transition: The Search for a Transformative Politics" (working paper #11, Trade Unions for Energy Democracy, 2018).
8. Bruce Podobnik, "Building the Clean Energy Movement: Future Possibilities in Historical Perspective," in *Sparking a Worldwide Energy Revolution: Social Struggles in the Transition to a Post-petrol World*, ed. Kolya Abramsky (Oakland, CA: AK Press, 2010), 74.
9. Ecosocialist critiques of the ruinous environmental impact of capitalism's innate competitive drive are increasingly widespread. For a representative example, see Joel Kovel, *The Enemy of Nature: The End of Capitalism or the End of the World?* (London: Zed, 2007).
10. To a certain extent this is the story of Orientalism: narratives of cultural supremacy were spun by Western European powers to justify their technological and military edge over once-threatening peer powers such as the Ottoman Empire. See Edward Said, *Orientalism* (New York: Vintage, 1979).
11. Timothy Mitchell, *Carbon Democracy: Political Power in the Age of Oil* (New York: Verso, 2011), 18.
12. Kenneth Pomerantz, *The Great Divergence: China, Europe, and the Making of the Modern Economy* (Princeton: Princeton University Press, 2001).
13. Nye, David E. *Consuming Power: A Social History of American Energies* (Cambridge, MA: MIT Press, 1999), 87.
14. Mitchell, *Carbon Democracy*, 12–42.
15. Rosa Luxemburg, *The Mass Strike* (Chicago: Bookmarks, 1986), 29–30.
16. Anne McClintock, *Imperial Leather: Race, Gender, and Sex in the Colonial Contest* (New York: Routledge, 1995), 76–79.
17. Robert K. Massie, *Dreadnought* (New York: Ballantine Books, 1991).
18. Mitchell, *Carbon Democracy*, 44.
19. Ibid., 29.
20. Ibid., 31.
21. Ibid., 46.
22. George Caffentzis, "A Discourse on Prophetic Method: Oil Crises and Political Economy, Past and Future," in Abramsky, *Sparking a Worldwide Energy Revolution*, 63.
23. Ervand Abrahamian, *The Coup: 1953, the CIA, and the Roots of Modern US-Iranian Relations* (New York: New Press, 2015).

24. Caffentzis, "Discourse on Prophetic Method," 63.

25. Robert Vitalis, *America's Kingdom: Mythmaking on the Saudi Oil Frontier* (New York: Verso, 2009), 88–120.

26. Abdelrahman Munif, *Cities of Salt* (New York: Vintage, 1989), 597.

27. Amitav Ghosh, "Petrofiction: The Oil Encounter and the Novel," *New Republic*, March 2, 1992, 29–34.

28. Rob Nixon, *Slow Violence and the Ecology of the Poor* (Cambridge, MA: Harvard University Press, 2012), 87.

29. Milton Viorst, "Desert Storm," *The Nation*, September 26, 2005.

30. On Khomeini's ideology as a form of third-world political populism, see Ervand Abrahamian, *Khomeinism: Essays on the Islamic Republic* (Berkeley: University of California Press, 1993).

31. Yaroslav Trifimov, *The Siege of Mecca: The 1979 Uprising at Islam's Holiest Shrine* (New York: Anchor Books, 2008).

32. Antonio Negri, *Marx Beyond Marx: Lessons on the Grundrisse* (Brooklyn, NY: Autonomedia, 1991).

33. For a discussion of this milieu, see George Katsiaficas, *The Subversion of Politics: European Autonomous Social Movements and the Decolonization of Everyday Life* (Oakland, CA: AK Press, 2006).

34. Ghosh, "Petrofiction," 33.

35. Matthew Huber, *Lifeblood: Oil, Freedom, and the Forces of Capital* (Minneapolis: University of Minnesota Press, 2013).

36. Cited in Huber, *Lifeblood*, 62.

37. Ibid., 62.

38. On the creation of surplus humanity, see Zygmunt Bauman, *Wasted Lives: Modernity and Its Outcastes* (Malden, MA: Blackwell, 2004).

39. George Caffentzis, "A Discourse on Prophetic Method: Oil Crises and Political Economy, Past and Future," in Abramsky, *Sparking a Worldwide Energy Revolution*, 60–71.

40. Jennifer Haigh, *Heat and Light* (New York: Ecco, 2016), 27.

41. Ibid., 385.

42. Mike Davis, *Prisoners of the American Dream: Politics and Economy in the History of the US Working Class* (New York: Verso, 2018).

43. Alex Kaufman, "The Surprising Origins of What Could Be the 'Medicare for All' of Climate Change," *HuffPost*, June 27, 2018.

3.

Gendering Petrofiction

Energy, Imperialism, and Social Reproduction

Sharae Deckard

As Egyptian novelist Nawal El Saadawi describes in her memoir, *Walking Through Fire*, she began writing the novel *Love in the Kingdom of Oil* during the Gulf War of 1990–91, shortly after her name had been placed by the "oil kings" on a "death-list" of writers accused of being heretics. She describes the novel's genesis as catalyzed by an epiphany as to the systemic nature of petromodernity: "Deep down inside me was the feeling that oil was the hidden actor behind all this. I gaze at the word 'oil,' underline it and this gives me a feeling of relief as though I have taken the first step towards naming the unknown force operating without a name."[1] Accordingly, her novel subjectivizes oil as an antagonist that structures the whole of daily reality for the unnamed female protagonist and depicts the exploitation of Arab women's bodies and the unvalued labor of their work of social reproduction as intimately bound up with petro-imperialism.

Sheena Wilson has recently called for a feminist approach that examines "how the inequities of race, class and gender are not only perpetuated in our current petroculture but also actively deployed as rhetorical strategies to literally and figuratively buoy and sustain existing power sources: oil and the neoliberal petro-state."[2] In particular, Wilson has criticized how mainstream Euro-American media instrumentalize rhetorics that invoke women's rights in order to rationalize colonial extraction politics and position female bodies as sites of spectacularized petro-politics while marginalizing the agency of women in acts of petro-resistance.[3] This essay seeks to extend this approach further by analyzing the gendering of world oil literature through two examples of feminist petrofiction from frontiers of oil extraction—Nawal El Saadawi's *Love in the Kingdom*, set in a petro-state resembling Saudi Arabia, and Laura Restrepo's *The Dark Bride*, set in Colombia. I argue that in these texts, questions of gender, of petrosexuality, and of the relation between energy and women's work in the realm of social reproduction are explicitly foregrounded in relation to oil imperialism.

El Saadawi's *Love in the Kingdom of Oil* offers a rebuttal to reductive imperialist narratives, such as the "clash of civilizations," that are mobilized to justify resource monopoly and incursion in the name of "modernization" or women's "freedom." Set in an unnamed Gulf petro-state, and prompted by El Saadawi's anger during the first invasion of Iraq, the novel rejects the crude attribution of the sexual oppression of women to Islamophobic accounts of Islam while at the same time fiercely criticizing the particular manifestation of capitalist patriarchy in the petro-state. Laura Restrepo's *The Dark Bride* (2001; *La novia oscura* 1999) is a historical novel set in the period from 1919 to 1951 when the US transnational corporation Tropical Oil Company—or "Troco," as it was colloquially known—operated in Colombia. The novel depicts the impact that the exploitation of oil fields in the province of Santander has on the local community of Tora. In particular, the book focuses on the relations between the male oil workers and the community of sex workers in a barrio called La Catunga. The climax is a fictional depiction of the 1948 oil worker strike, in which sex workers joined; as such, the novel could be considered a feminist strike epic, which crucially foregrounds the subjective experiences of women in acts of energy resistance rather than focusing exclusively on the masculinized oil proletariat.

My approach in conjoining these texts follows recent calls by critics for a world-literary approach to petrofiction. Thus, Imre Szeman has outlined a critical approach to "world energy literature" that includes "energy inequality as an axis of analysis," seeking to discern how "inequalities, not only of financial power or fictitious debt, but also of energy flows, forms and capacities" are figured differently between cores and peripheries.[4] Similarly, Graeme Macdonald has argued for the "necessary 'worlding' of petrofiction and resource texts" to read how literary forms mediate the "carbon flows, exchanges, relations and circulations" of the "fuel-ecological world-system."[5]

Both critics suggest that the "occlusion argument" about oil's alleged invisibility widely popularized by Amitav Ghosh's famous essay on petrofiction (1992) appears in a different light when we examine global petrofictions that depict the frontiers of oil's extraction. For Szeman, the periphery provides "imaginative openings" that counter "the disassociation that pushes oil to the background in the global north."[6] Likewise, Macdonald asserts that the critical insistence on the invisibility of oil "increasingly stands alongside recognition of its spectacular violence and the huge material impositions that have accompanied its terraforming of territories and reorganization and publics across the world," and he notes that "for the many extraction sites on the (semi-)periphery of the world-system—*and* within cultural production from those areas—oil is or has been overtly visible, even if it is subsequently made 'unseen,' either by privatization, securitization and military enforcement *or* by its mediated mystification."[7] The spaces might be concealed to those in cores, but to those living in such sites of extraction, oil is not so much "hidden" away as openly "taken."

I am particularly interested in what comparing literature that represents nodes of extraction and export—those enclaves in which crude oil is drained away, to be produced and consumed elsewhere—might reveal about how particular socioecological relations of gender and sexuality congeal around oil frontiers in the periphery and about how the historical emergence of these constellations might be mediated differently in aesthetics. Indeed, although El Saadawi's essay initially uses the language of occlusion when referring to oil as a "hidden actor," she swiftly exchanges it for a more subjectivized depiction of oil's agency: "In my novel, oil is the hero. The island in the story is floating on a sea of oil, completely under the control of an oil consortium."[8] In *Love,* oil's local presence is far from concealed but rather is explicitly, even excessively, legible. There is scarcely a page that does not mention oil by name or is not stained by the black rains, the seeping floods, and the heavy jars full of dark liquid that oppress the protagonist. Both texts are saturated by a "crude materialism" in which oil is explicitly depicted as the form of capital "that bulks out and inhabits"[9] the local places they represent, violently transforming the quality of their environments, indelibly shaping the psychosocial dynamics of their characters, and underwriting daily relations with a sense of immanent disaster. However, *Love* is an ostentatiously irrealist novel, hallucinatory in its aesthetics, in contrast to the more realist aesthetics of *Dark Bride.* Part of this difference in mode can be attributed to formal pressures on representation and part to political intent and the genealogy of the contextual materials on which the novels draw.

Love's surrealism is more akin to that of oil modernism, whose extremity derives both from its thematic focus on the massive shocks of social transformation engendered by uneven oil modernization and from its narrative attempt to subjectivize oil as a historical force. *Dark Bride*, by contrast, is a "historical" novel with documentary and socialist intent, directly influenced by interviews conducted by Restrepo, who visited the Magdalena Medio in the 1990s as a journalist investigating the criminal activities of a gasoline cartel and encountered the survivors of the famous 1940s oil strikes. In *Dark Bride*, the gender relations that constellate around the early twentieth-century frontier are foregrounded, but oil as energy, as the logic of late fossil capital, is not itself formally subjectivized as a "hero." Neither is there the same dizzying sense of phantasmagoria that accompanies the in situ representation of the petro-state built on "anomalous energy surplus,"[10] although the figure of Sayonara herself, the dark bride, is imbued with a mystery and compulsive attraction associated with the fetish of oil. With the clarity of its retrospective vantage on an earlier stage of oil imperialism, the text in *Dark Bride* represents local place with an intensely concrete specificity, mapping a "functional human point of view" on petromodernity by linking "the abstraction of social 'totalities' to domestic intimacies."[11] The two novels share a central focus, however, on the gendering of labor in relation to the intimate sphere of the domestic household—the hidden abode of social reproduction. Each articulates in its own way a feminist critique of how patriarchy and petrocapitalism have worked in tandem. In what follows, I will trace the relation between energy, women's work, and social reproduction and explore how the two texts offer insights into the ways that constellations of gender and sexuality in these two different regions and periods underlie the extraction of energy as surplus value but also recuperate modes of resistance and bring back to memory the justice strategies pursued by women.

Energy, Women's Work, and Social Reproduction

In social reproduction theory, "reproductive work" encompasses the whole complex of social relations, institutions, and activities that produce and reproduce labor power for capitalism—from domestic work, sex work, child-rearing, and care work to the unpaid forms of subsistence farming and resource gathering (as in water or biomass collection) included under the rubric of housework in many parts of the Global South.[12] George Caffentzis's classic essay "The Work/Energy Crisis and the Apocalypse" offers a useful starting analysis of how the role of energy in the Keynesian crisis of accumulation was bound up with struggles against work originating not only in factories but in conjunction with the resistance of unwaged workers in the kitchens, households, and bedrooms of the world. In the mid-1970s, it became advantageous for capital to increase the cost of reproduction in order to attack wages through administered increases in the prices of oil.[13] This attack on working-class wages shifted the costs of reproduction even further onto the unwaged work performed mainly by women. As such, energy is intimately bound up with the housewifization of labor, the systemic usurpation by the capitalist patriarchy of subsistence work without credit for the economic benefits it provides. Without this mass of unpaid work, both physical and affective and usually performed by women, capitalist societies would not be able to produce new waged workers to work in the extraction, production, refining, or distribution of energy commodities.

Social reproduction theory has rightfully attained increasing prominence in feminist critiques of capitalism, but as Kolya Abramsky notes, surprisingly little has yet been written connecting energy with social reproduction, much less differentiating the particular modes of energy entangled with it. These include energy as a "means of reproduction/subsistence," as "resources" predominantly located on and extracted from rural land, and "as unwaged labor in the non-commercial energy sector." Within these modes, the question of the gendering of the unpaid work that enables energy extraction and resistance to energy regimes on the terrain of social reproduction is key.[14] The profits of the commodity frontier of oil extraction are bound up with the appropriation of the unpaid work/energy of women in the realm of subsistence, turning on the devaluing—or what Raj Patel and Jason W. Moore would call the "cheapening"—of their labor.[15] I turn now to examining the relation between oil and social reproduction, contending that it is a crucial task for feminist petro-critics to consider how these relations are differently articulated according to period, region, and the particularities of local culture but might also share similarities, particularly in terms of the tropes or aesthetics adopted to figure them in global literary representations. Taking up the thread of domestic intimacies, I begin by examining how the depiction of "oil wives" is used in these texts to represent the phenomenon of housewifization.

Matthew Huber demonstrates in *Lifeblood* how oil was framed in the media in 1950s America as essential to reducing the drudgery of household labor: from remaking the everyday life of the housewife and her family by improving the practices of cooking, cleaning, shopping, childcare, beautification, and sleeping, to ushering in the age of easy modern living through the assistance of new electronic devices, plastics-derived products, cosmetics, and foods. Discussing an ad that lauds oil as "Mother's little helper," Huber observes that "oil was framed as the thread holding together this seemingly disconnected series of practices of social reproduction. Without oil, life, family, and everything else *could not be reproduced*."[16] He simultaneously observes, however, that such petroleum advertisements tended to portray exclusively white images of femininity, targeting only certain racial and economic demographics with privatized geographies of oil wealth.[17] The "good life" based on the astounding energy surplus of US oil and generalized mass consumption in Fordist America was available only to a certain kind of affluent suburban housewife.

Contrast the advertisement of oil as housewife's helper, enabling a "life in *opposition to work*"[18] to the image of the "wife of oil" that appears in El Saadawi's *Love in the Kingdom*. Her life is nothing *but* work in the realm of the household.[19] The unnamed archaeologist, in search of archaeological evidence of a matriarchal culture that pre-dates patriarchal religion, arrives on a mythical island in a sea of oil. She soon finds herself the unwilling hostage of a local man, who in the book's surreal dream logic transmogrifies into a "second" husband. This "husband," whom she describes as "like a child . . . who could neither feed himself nor get himself something to drink,"[20] performs a petro-masculinity that is innately insecure, rooted in the hollow promises of oil plenitude. He is wholly dependent on the woman's labor for his own subsistence, both physical and affective, and demands that she assuage the insecurity that perpetually undermines his proprietarian identity: "The gushing of the oil was eating away at the wall, and jealousy was eating away a bit of his flesh under the twisted rib."[21] He physically coerces her to perform chores stereotyped as women's work: cooking dinner, fetching water, servicing his sexual appetites, and carrying jars of black liquid to "the company," the unnamed transnational oil corporation run by a white American man. The protagonist is disciplined by the physical threat of hard violence, by the ideological coercion implicit in the other women's surveillance of her behavior, and by the eerie omnipresence of oil itself, which relentlessly soaks up her strength and drains her of dissent, so that she becomes a zombie-like automaton:

> The oil had soaked up her strength, and the man had finished filling the jar. He stood waiting for her to move. He began to stare at her for a long time and then raised his arm upwards. She had the idea of resisting, of returning blow for blow. But her arm

> remained stuck to her side. Perhaps it belonged to a wife of oil that made things stick together. . . . Or perhaps the unexpected blow had deprived her of her will and made her kneel down like a camel. He placed the wase on her head and the jar on top. It was up to her to walk with the ladies to the company. Each one of them gave her word, just for the sake of chattering on the way to the company. "Do you understand? He will not beat you if you continue to work."[22]

The heavy jars of oil that she is driven to carry have multiple connotations. They allude to the feminization of water provision in many cultures throughout the Global South, where the taxing manual labor of fetching water is considered women's work. In their strange inversion of oil for water, they also signify the gendered hydropolitics of what Toby Jones has called the "Saudi alchemy" of "turning water into oil, oil into water."[23] This phrase refers to the dependence of oil abundance in Saudi Arabia on a hydraulic regime that exhausts aquifers in order to pressurize local oil fields, pumping in water to force crude oil out, while at the same time using imported water and wheat bought with oil rents as patronage to secure the loyalty of local elites and pacify dissent. The novel echoes this alchemy in the paltry ration of "oil drops" allocated by the monarchy to each citizen to satisfy their thirst in place of water. They become addicted to these drops, drinking, as it were, their own small portion of rents in exchange for their political subservience to the king. Hydrological crisis is figured incessantly throughout the text, whether in the woman's unslakable thirst or the constant substitution of the semiotics of oil for water: the black lake, the oil falling from the sky instead of rain, the dark liquid running from taps.

Finally, the jars of oil suggest the extent to which all forms of labor—even those unwaged forms outside the formal sector—become imbricated in an oil economy dominated by foreign interests and constellated within a petroculture that shapes every aspect of social relations, so that even unpaid women's work services the immense oil profits of the company. The novel's extraordinary use of analepsis and prolepsis, incessantly doubling back and moving forward between the woman's past and present until all chronology is blurred, can be read as figuring the temporality of oil's perpetual circulation and the cyclical periodicity of boom and bust. But it also acts as commentary on the complex operation of patriarchy in tandem with petro-capitalism, as all the male figures of authority blend into a continuum of multiple "husbands"—her boss, her father, the policeman investigating her disappearance, her first husband, her second—thus representing the legal, economic, and familial social institutions that work together to dictate her behavior and police her subjectivity within a patriarchal conception of gender norms and that operate to service the needs of a petro-state that is itself subservient to the larger interstate dynamics of petro-imperialism. When accused of blasphemy, the woman is quick to assert that her oppression is rooted not in her own (heterodox) faith but rather in the material circumstances of her economic exploitation, which the patriarchal form of religion dominant in the state is used to rationalize: "[Her heart] was large enough to embrace a faith bigger than his faith, a faith that also included the ancient gods—and the goddesses. But what was the relation of gods with money? . . . She carried a jar on her head in an official oil company."[24]

The novel thus offers a more complex mediation of how oil production intersects with the gendering of the labor force than is available from many political theorists. Michael L. Ross's controversial article "Oil, Islam, and Women" is interesting in this regard, arguing that oil production in Middle Eastern societies fundamentally rearranges the composition of labor and squeezes women out of the formal economy, leaving oil-producing states with "atypically strong patriarchal norms, laws, and political institutions."[25] Ross foregrounds the importance of political economy and usefully rejects the "clash of civilizations" thesis that Islam is to blame for what he teleologically frames as "lack of progress towards gender equality in the Middle East,"[26] but he perpetrates monolithic claims about non-Western oil-producing societies as rentier states, combined with an unsophisticated analysis of patriarchy and a liberal embrace of capitalist development.

In contrast, El Saadawi's analysis of the relation between women's oppression and petro-imperialism rejects the ideological attribution by "Western imperialist thinkers" of the "problems of Arab women [to] the substance and values of Islam" as well as to their failure to understand so-called oil modernization not as a process of *development* but rather of inequality, exploitation, and "plunder."[27] In her essays, she contends that the purpose of oil-driven development is not emancipation but forms of technological advance aimed at more rapid extraction, "quicker and bigger profits, and more effective and efficient ways of pumping out oil from under the shifting desert sands or the depths of ocean beds." The development of extractive infrastructures is subsumed under the imperative that "such resources must continue to serve the interests of international capitalism and the multinational giants that still rule over a large part of the world."[28] Petro-imperialism as described by El Saadawi works in conjunction with patriarchy to perpetuate a gendered division of labor and preserve the political power and grotesque riches of a handful of national elites who serve the larger interests of hegemonic capitalist cores.[29]

This recalls Heather Turcotte's crucial insights in the context of oil imperialism in the Niger Delta that petro-violence materializes in "multiple forms that have always been gendered and systemically violent" and that gender violence is not some mere *effect* of petro-violence to be understood through simplistic causality but rather is the necessary condition for "a larger political economy of violence that creates the conditions fostering and facilitating petro-politics in the first place."[30] These conditions are rooted in the gendered history of colonial domination, where colonial administrations frequently utilized sexual and gender violence as "a means of securing productive and reproductive resources that include bodies, labor, and land as well as oil" and where "the proletarianization of men rested on the domestication of women through the physical segregation of public and private places."[31]

Throughout *Love in the Kingdom*, the private space of the house is repeatedly invaded by what Layla Hendow calls a "pornography of oil."[32] A "black dampness with a pungent smell" soaks the bed sheets, "black liquid" pours down profusely "with a sound like a waterfall" into the kitchen, "tiny black particles" fill up her eyes and ears, "threads of black liquid" push their way in under the door, and the whole house seems to rock on a sea of oil "as if it were a boat."[33] This oil-suffused household is neither a paradise of petrolic good life nor a domestic refuge from the toil of work but rather an infernal nightmare. Within the novel's allegorical society, all women are wives of oil, and the hidden abode of reproduction is cast as a horror show in which oil suffuses every aspect of daily life down to the walls, wherein the housewife is doomed to perform eerily repetitive chores, evicted from the sphere of formal labor.

Even the machine appears differently here than in Huber's account of midcentury American life, as something monstrous rather than wondrous. The apparition of a "new, oil-powered typewriter" metamorphoses into a masculinist fantasy of a machine that can "carry jars on four rubber feet, and walk by oil power" and replace women altogether in the sphere of reproduction and care. This machine is an uncanny counterpart of the fetishistic personification of oil as "Mother's little helper"; that it arises in response to fears that women's demands for wages will cause "an insane rise in prices" suggests a correlation between the ideological interests of patriarchy and the economic realm, such that struggles around reproduction exert pressure on capitalist accumulation. The husband's fantasy that he could carry his wife on his back "as if she was a lamb, and place her on the scales, and with the price he received he would buy that new machine"[34] is not just a ludicrous dream turning on the misogynist erasure of women from a society through the magic of oil; rather, the threat of replacing women with cheap energy serves to maintain the low value, or devaluing, of cheap labor and cheap care.

The use of housewifization and the instrumentalization of the nuclear family to secure a particular regime of extraction appears in a different configuration in Restrepo's *Dark Bride*, in which the titular *novia oscura* could be understood as referring both to Sayonara—the *mestiza* prostitute—and to oil itself. The novel opens by describing the constellation of gender relations that has emerged to support the labor of oil extraction on the frontier during the early twentieth century: "Back then Tora was distinguished in the great vastness of the outside world as the city of the three p's: *putas*, *plata*, and *petróleo*, that is, whores, money, and oil. *Petróleo, plata*, and *putas*."[35] In 1920s–40s Tora, it is not initially in the realm of the household that women's labor is most significant but in the realm of sex and care work. As the *petrolero* Sacramento comments, the promise of recreational escape in the brothels of La Catunga is what fuels the transnational force of male oil workers in *petrolero* camps in the interior: "Just for that we would break our backs working in the cruel jungle, the four hundred workers of Campo 26. Thinking of that sweetness, we withstood the rigors of Tropical Oil."[36] In this gendered division of labor, the city of brothels run by women is a necessary counterpart to the realm of international proletarianized male labor on the oil fields: "The men from all the camps in the world, from the oil wells of Infantas to the vast fuel deposits of Iraq, passed religiously through the streets of sin in La Catunga, as if coming to fulfil a promise, because it was the heart and sanctuary of the extensive oil labyrinth."[37]

The majority of the oil workers do not shame the women for their profession but call them only "las mujeres," since "in the oil world, *amor de café* was the only recognized form of love."[38] In turn, the *madrinas* view their trade as comparable to any other waged labor. As Todos los Santos puts it, refusing the moral dichotomy of wife versus whore, "I have always believed that a *puta* can have a life that was just as clean as any decent housewife, or as corrupted as any indecent housewife."[39] But the novel is careful to record the different extents to which their experiences of sexual commodification affect them emotionally and the pernicious ways in which the *barrio* is divided along class and ethnic lines, so that *pipatónas*, Indigenous women from the interior forced into prostitution after their land is expropriated for oil fields, earn the least, occupy the poorest brothels, and are the targets of more sexual violence, even if they also tend to refuse to internalize Catholic guilt. *Dark Bride* subverts the generic tendency of Latin American novels such as Mario Vargas Llosa's *The Green House* to use the brothel as a microcosm of power and national identity while privileging a "dominant masculine viewpoint" and producing grossly objectified characterizations of women.[40]

The particular form of petro-masculinity characterized by Restrepo's *petroleros*—which is very different from the parasitic relation between oil wives and husbands in the petro-state described in El Saadawi's novel—is rooted in a sense of male virility linked to manual labor and the vitality with which the men operate the rigs and drill the earth and to their financial ability to purchase the love of *las mujeres* with their Troco wages. Sacramento describes this archetype of the hirsute, manly oil worker in terms that evoke Cara Daggett's description of the "hard masculinity"[41] associated with fossil-fueled sexuality: "An oil worker was proud of arriving in La Catunga looking tough, tan, hairy and bearded. But clean and smelling fine, wearing leather boots and a white shirt, with a good gold watch, necklace, and ring to show off his salary. And always, as if it were a medal, his company ID visible on his lapel."[42] This affirmation of masculinity through the labor of oil extraction, however, is repeatedly troubled throughout the novel.

During the rice strike, the sense of pride derived from their wages is transformed into rage and exhaustion as the workers awaken to the conditions of their exploitation: the "excessive work" that burns into their muscles and the "deafening noise of the machines" that clutters their skulls and crowds out their thoughts.[43] For the character Payanés, the identification of masculinity with the labor of oil extraction—specifically, his work as a *cuñero*—takes on pathological dimensions when he fetishizes Skinny Emilia, the machine he operates: "He talked about that tower as if rather than a framework of steel it were a woman, exposed and solitary."[44] This has both interpersonal and political consequences: his "love" of skinny Emilia causes him to block the strikers from smashing it to prevent scab labor, with the result that the strike is more swiftly broken by the importation of outside labor, and he is unable to form a long-term relationship with Sayonara, his lover, abandoning her after the strike in search of a new rig.

Sacramento's version of petro-masculinity is tormented by his internalization of shame at being an *hijo de la Catunga*, the son of a sex worker, abandoned to a Franciscan orphanage school. Instead of embracing the morality of the community, in which "the normal thing was to be a *puta*, and to be an *hijo* de puta—the son of a *puta*—was the logical and painless consequence," he is tortured by a Catholic ethos that fills "his soul with a horror of the sins of the flesh and with a visceral mistrust of women."[45] This fault line of guilt intermingled with misogyny prompts Sacramento to become a scab by signing up for a subsidized house and a company-officiated marriage to Sayonara, whom he desires to "save" from prostitution. Their marriage subsequently becomes a suffocating trap, in which the free-spirited Sayonara is transformed into an effigy of wifely purity.

Their doomed marriage serves to generically subvert the Latin American genre of historical romance—in which "indio" or "mestiza" women are happily married off at the plot's conclusion in a mode of bourgeois reconciliation—by deliberately exposing the marriage as oppressive, founded on hierarchies that cannot be reconciled but only overturned, and then by breaking it off, to allow Sayonara to pursue her own path once more. At the same time, the deflation of the central love match can be read through the prism of narrative energetics as the formal expression of the collapsing social energies of the larger strike after it is repressed. Furthermore, the failed marriage plot also enables the critique of how oil capital works through patriarchy, Catholicism, and a heteronormative conception of the nuclear family to institute a new form of gender relations that recontains the energies of the strike and prevents further dissent. The oil company realizes that

> to have rootless men piled up in barracks with a hammock and a single change of clothes as their only belongings and with a *puta* as their only love, or in other words with everything to gain and nothing to lose, was to be confronted by bitter enemies that were impossible to manage. On the other hand, a man with a house, a wife, and children, which sizable burden the company helped him to support, would think twice before risking his job to join the fight.[46]

This transition from *petrolero* camps to the modern form of a company town is accomplished through violent tactics, as the "four sources of power" in the region—Troco, the Fourth Brigade, the mayor, and the Church—work in "synchronization" to "cleanse" La Catunga and replace it with a suburban neighborhood.[47] These powers collude in beating and shaving the prostitutes, in forced clearances of the red-light district, in internment of fifty women "in a detention camp with barbed wire and military guards" run by the Fourth Brigade, and in the demonization of "sick" *putas* by an imported troop of Franciscan monks.[48] Thus, the novel's original trinity of *putas, petroleros,* and *petroleo* is reconfigured after the strike (which historically took place in 1948) to usher in a new regime of gender relations, in which unwaged housewives rather than sex workers are expected to take on a central role in providing the reproductive work that guarantees the stability of its labor force. The novel chronicles the historical pivot in midcentury to housewifization, enabled by the modernization and standardization of the company oil town whose suburbs promote a clear division of labor organized around the nuclear family and serve to contain any prospect of striking energies.

The depiction of this crucial historical transition is juxtaposed with the contemporary perspective of the frame narrative, focalized through the reporter's view in the 1990s of the era of late fossil capital, in which decades of the slow violence of pollution together with cartel conflicts over gasoline and narco-territory have reduced the town to a toxic wastescape:

> The formerly great Río de la Magdalena seemed to me like a long absence: slow, black, full of dredging boats—could those brown monsters that sank their feet in the water be dredgers?—and other metallic and orthopaedic apparatuses that turned it into an extension of the refinery, which spread across the opposite bank, rusting the night sky with the perpetual combustion pouring from its tall smoke-stacks. An incongruent smell, feminine and sweet, came from those iron pipes. Don Pitula, the taxi driver who guided me around Tora—and who had worked as a welder at the refinery for twenty-five years—had told me that afternoon that the perfumed smoke came from a factory that made aromatics, where they processed petroleum, into shampoo, facial creams, and other cosmetics. "The factory that smells the best is the most poisonous," he told me. "Working there is like signing a death sentence."[49]

This gothic passage exposes the socioecological violence underlying the production of the petrochemical cosmetics through which contemporary petrofemininity is expressed by the global consumers of periphery's exports. By juxtaposing the factory in which they are produced and the oil refinery on which they rely, the narrative points to the intersection between energy and the changing composition of the labor force brought about by the feminization of the proletariat in the neoliberal era, specifically the "cheap" work of the transnational force of female factory workers in the neoliberal era, who are treated as disposable bodies to be exposed to toxic working conditions without recompense. The novel's act of totalizing retrospect by juxtaposing this deathly present with the history of past energy struggles can be understood as driven by the twin imperatives to understand the historical causes for the contemporary nexus of oil extraction, environmental degradation, and gender violence in the Magdalena Medio region and to recuperate those social conjunctures that once intimated alternative futurities.

Petro-protest and the Terrain of Social Reproduction

Even though the plot of *Dark Bride* culminates in the eradication of the solidarity established between *mujeres* and oil workers, it retains its importance as a narrative of protest that foregrounds the potential for energy struggles to develop over social reproduction. Salar Mohandesi and Emma Teitelman have recently called for a history of class formation that moves beyond a confined focus on male factory workers to the terrain of social reproduction, understood as "not only a site of struggle, but a potential site of class formation" on which "actions over social reproduction could not only trigger struggles elsewhere, but fuse them together."[50] In this spirit, I will conclude by exploring how struggles over social reproduction in the context of petroculture are represented in these texts.

While domination can feel totalizing due to the formal logic of recurrence in *Love in the Kingdom*, the novel is not merely a chronicle of oppression but is instead punctuated insistently with expressions of resistance, even if these are subsequently recontained. The researcher repeatedly attempts to mobilize the women around her to action, so that her "husband" admonishes her: "Oil can be friendly towards us if we surrender to it, but you never stop resisting it."[51] In one instance, the researcher attempts to spur the other women carrying jars into a direct confrontation with the oil consortium, asking, "Isn't there anybody to resist the oil? Don't you ever think of solidarity?"[52] In another striking passage, the researcher watches the women sing and dance in a moment of collective resistance:

> Women were dancing in a circle, their feet advancing to the same beat. The singing rose to heaven with the dust.
>
> "Is it our fate to carry on our heads . . ."
>
> "Jars of oil for ever?"
>
> "No, sister! No, sister!"
>
> "It's not our fate! It's not our fate!"
>
> It was astonishing to see the wisps of light in the darkness. To discover the connection between fate and oil. Her body appeared to her like the wall, planted firm in the face of the storm. Nothing could topple her.[53]

Here, the individual's sense of personal agency is revitalized in connection with the collective action of the other jar carriers, even though she is subsequently silenced once more. As such, the fact that the narrative ultimately does not deliver freedom for its protagonist, only a circulation back to the beginning, could be read not as fatalism but as an expression of El Saadawi's own conception of political struggle: "If you're an individual, they can smash you, silence you, but if you're doing collective work, it's much more difficult for them."[54] Neither are expressions of struggle in the narrative confined to abstract defiance; rather, they are transmuted into repeated demands for wages and recognition of women's reproductive labor: "Haven't we demanded our wages before?" "Yes, we have." "Then we must stop demanding them and take them with our own hands."[55] The focus here is not on the female body as a site of spectacularized politics but rather on the demands of women for economic justice.

In *Dark Bride*, the oil workers' first uprising is called the "rice strike" because it is triggered by their disgust with the inedible "rice ball" ration and their frustration at their working conditions and inability to form families. The women's first political action is catalyzed when Sayonara protests at the venereal disease clinic, where the women are subjected to abusive examinations and expected to pay for a weekly card licensing them as "fit" to work, even though they do not receive any treatment. A riot ensues against what they rightly perceive as an illegal tax on their labor as the "mutinous *putas*" set the dispensary on fire.[56] Although the army is brought in to suppress them, they are successful in closing the clinic and ending the tax. Later, when the rice strike swells into a citywide demonstration, the women join in.

> The *prostitutas* of La Catunga went on strike with legs crossed in solidarity with the *petroleros* and stayed out of the café. They traded dangly earrings and diadems for

> red rags that they tied around their heads and took to the streets, along with the general populace, to participate in the manifestations that arose on every street corner and to join protests and massive acts of resistance in support of the list of demands. And, out of an extra sense of civic concern, they demanded an aqueduct and sewers in the neighborhoods of Tora, which were burning with thirst and drought.[57]

Thus, Restrepo depicts struggles over sex work, the affordability and quality of food, accessibility to clean water, and waste disposal—all central issues of social reproduction—as the terrain that catalyzes and fuses the struggles of both women workers and oil workers.

The novel has been called Zola-esque, displaying thematic similarities to the depiction of miners' strikes in *Germinal* and to the world of prostitutes in *Nana*. However, the novel rejects the "straitjacket" of naturalism in order to foreground the "herstory" of the women's resistance.[58] Rather than being condemned to a lingering death like Nana, Sayonara gains political agency; indeed, as she operates the mimeograph during the strike, she and her fellow strikers are described as experiencing "a moment that must have made them feel like protagonists not only in their own personal drama but also in the history of their nation." She thus transcends the limits of her inscription within the doomed romance narrative to become a protagonist of the strike epic.[59] Women are not fetishized as victims, though they are subjected to different forms of violence. Instead they are foregrounded as political actors: intelligent, resourceful, and capable of both spontaneous action and political organization. Their "justice strategies" might not be "petro-protests"[60] in the sense of being explicitly directed against the ending of oil as the energy regime underlying petromodernity, but they are energy resistance in the sense of struggling against the particular ecological regime in which socionatural relations of gender, class, and energy have been organized. As such, the novel could be considered a feminist reprise to that other famous Colombian strike novel, *100 Years of Solitude*, which foregrounds women's involvement on the terrain of social reproduction in the historic "rice strike" against the Tropical Oil Company in the interior.

Near the end of the novel, Frank Brasco, a manager who switches sides and joins the strike, describes the protest as producing an extraordinary transindividual affect: "The cocktail of collective enthusiasm, solidarity, and fear was so explosive . . . that you could say that we were all in love with everyone."[61] For the frame narrator, speaking from the late 1990s, this sense of "communal eroticism" is crucial to recuperate because of how it opens up the imagination of "a dazzling succession of futures that we Colombians have never experienced"—that is, because it recovers a prospect of political futurity that seems foreclosed in the neoliberal present.[62] In conclusion, I would suggest that libidinal figurations of futurity such as these might lead petro-critics to envisage feminist energy futures and transitions, moving beyond merely improving the conditions of access to food, water, and energy and instead imagining how the terrain of reproduction might trigger forms of social transformation whose desire is for more emancipatory organizations of nature.

Notes

1. Nawal El Saadawi, *Walking Through Fire: A Life of Nawal El Saadawi*, trans. Sherif Hetata (London: Zed Books, 2002), 16.

2. Sheena Wilson, "Gender," in *Fuelling Culture: 101 Words for Energy and Environment*, ed. Imre Szeman, Jennifer Wenzel, and Patricia Yaeger (New York: Fordham University Press, 2017), 177.

3. Ibid.

4. Imre Szeman, "Conjectures on World Energy Literature: Or, What Is Petroculture?," *Journal of Postcolonial Writing* 53, no. 3 (2017): 281.

5. Graeme Macdonald, "'Monstrous Transformer': Petrofiction and World Literature," *Journal of Postcolonial Writing* 53, no. 3 (2017): 291, 292.

6. Szeman, "Conjectures," 284.

7. Macdonald, "Monstrous Transformer," 293.

8. El Saadawi, *Walking*, 17.

9. Stephanie LeMenager, *Living Oil: Petroleum Culture in the American Century* (Oxford: Oxford University Press, 2014), 13.

10. Szeman, "Conjectures," 286.

11. LeMenager, *Living Oil*, 133.

12. Matthew Carlin and Silvia Federici, "The Exploitation of Women, Social Reproduction, and the Struggle Against Global Capital," *Theory and Event* 17, no. 3 (2013): para. 13.

13. George Caffentzis, "The Work/Energy Crisis and the Apocalypse," in *Midnight Oil: Work, Energy, and War, 1973–1992*, ed. Midnight Notes Collective (New York: Autonomedia, 1992), 215–71.

14. Kolya Abramsky, "Energy and Social Reproduction," *The Commoner* 13 (2016): 338.

15. Raj Patel and Jason W. Moore, *A History of the World in Seven Cheap Things* (Oakland: University of California Press, 2017), 111.

16. Matthew T. Huber, *Lifeblood: Oil, Freedom and the Forces of Capital* (Minneapolis: University of Minnesota Press, 2013), 83.

17. Ibid., 85.

18. Ibid.

19. Nawal El Saadawi, *Love in the Kingdom of Oil*, trans. by Basil Hatim and Malcolm Williams (1993; London: Saqi Books, 2001).

20. Ibid., 28.

21. Ibid., 132.

22. Ibid., 72–73.

23. Toby Craig Jones, *Desert Kingdom: How Oil and Water Forged Modern Saudi Arabia* (Cambridge, MA: Harvard University Press, 2010), 24.

24. Ibid., 81.

25. Michael L. Ross, "Oil, Women and Islam," *American Political Science Review* 102, no. 1 (2008): 107.

26. Ibid.

27. Nawal El Saadawi, preface to *The Hidden Face of Eve*, in *The Essential Nawal El Saadawi: A Reader*, ed. Adele Newson-Horst (London: Zed Books, 2013), 44.

28. Ibid.

29. While El Saadawi seeks to emphasize a critique of petro-imperialism rather than to advocate a version of rentier state theory in her political writing, it is important to note that her novel could be viewed as problematic in the way that it represents an unnamed, abstracted Gulf state. In his astute reading of petro-drama from Kuwait, Faisal Adel Hamadah has usefully cautioned against the danger of promulgating "neo-Orientalist caricatures" of the Gulf states as a homogenous bloc of "rentier states" whose "populations are all happy to live off wealth" or in which all women are uniformly "repressed" and the particularities of different political movements, historical forms of resistance, and state formations as well as the hierarchical relations between the states are erased. The allegorical form of *Love in the Kingdom* does tend to present its unnamed petro-state in rather monolithic terms, and even if its critique seems implicitly aimed toward Saudi Arabia, whose role in supporting US imperial hegemony is particularly significant, this is not explicitly concretized within the narrative.

30. Heather Turcotte, "Contextualizing Petro-sexual Politics," *Alternatives: Global, Local, Political* 36, no. 3 (2011): 200.

31. Ibid., 206.

32. Layla Hendow, "Oil and Women: Invisibility as Power in Nawal El-Saadawi's *Love in the Kingdom of Oil*," in *Seen and Unseen: Visual Cultures of Imperialism*, ed. Sanaz Fotouhi and Esmail Zeiny (London: Brill, 2017), 90.

33. El Saadawi, *Love in the Kingdom*, 9, 29, 20, 22.

34. Ibid., 6, 98, 96.

35. Laura Restrepo, *The Dark Bride*, trans. Stephen A. Lytle (1999; New York: Black Swan, 2001), 8.

36. El Saadawi, *Love in the Kingdom*, 7.

37. Ibid., 80.

38. Ibid., 8.

39. Ibid., 79.

40. Kate Averis, "Laura Restrepo (1950)," in *The Contemporary Spanish American Novel: Bolaño and After*, ed. Will H. Corral, Nicholas Birns, and Juan E. De Castro (New York: Bloomsbury, 2013), 255.

41. Cara Daggett, "Petro-masculinity: Fossil Fuels and Authoritarian Desire," *Millennium: Journal of American Studies* (2018): 1–20.

42. Restrepo, *The Dark Bride*, 8.

43. Ibid., 231.

44. Ibid., 264.

45. Ibid., 25.

46. Ibid., 276.

47. Ibid., 358.

48. Ibid., 352.

49. Ibid., 112.

50. Salar Mohandesi and Emma Teitelman, "Without Reserves," in *Social Reproduction Theory: Remapping Class, Recentering Oppression*, ed. Tithi Bhattacharya (London: Pluto, 2017), 47.

51. El Saadawi, *Love in the Kingdom*, 90.

52. Ibid., 75–76.

53. Ibid., 85.

54. Nawal el Saadawi, cited in Peter Hitchcock, Nawal El Saadawi, and Sherif Hetata, "Living the Struggle," *Transition* 1 (1993): 172.

55. El Saadawi, *Love in the Kingdom*, 130.

56. Restrepo, *The Dark Bride*, 71.

57. Ibid., 273.

58. Stephen M. Hart, *A Companion to Latin American Literature* (Rochester, NY: Tamesis, 2007), 259.

59. El Saadawi, *Love in the Kingdom*, 272.

60. Turcotte, "Contextualizing," 209–10.

61. El Saadawi, *Love in the Kingdom*, 272.

62. Ibid.

4.

Petrofeminism
Love in the Age of Oil

Helen Kapstein

Don't we all in the end write about love? All literature is about love. When men do it, it's a political comment on human relations. When women do it, it's just a love story. So, although I wanted to do much more than a love story, a part of me wants to push back against the idea that love stories are not important. I wanted to use a love story to talk about other things. But really in the end, it's just a love story.
—Chimamanda Ngozi Adichie to Emma Brockes, interview in *The Guardian*

When writing *Americanah*, Chimamanda Adichie self-consciously "wanted to use a love story to talk about other things," but love stories are always about other things.[1] In Nigerian love stories—whether mainstream romance novels from the South or the "love literature" of the Islamic North—one of those other things is oil. When Peter Hitchcock refers to "oil's generative law," he means that it is "everywhere and obvious[;] it must be opaque or otherwise fantastic."[2] To expand, oil not only generates but regenerates and is gendered. Here, I read Nigerian romance fiction for what it says about gendered relations to oil, showing that oil is indeed everywhere and making that ubiquity obvious.[3] In the course of writing about how Nigerian short stories sabotage Big Oil's narrative, I discovered that romance is the most widely read genre in Nigeria and that 91 percent of romance book buyers are female.[4] My hypothesis for this study was that, given the total saturation of everyday Nigerian life with oil politics, its tensions and debates must inevitably arise in the country's fiction. As it turns out, while the short stories explicitly call out the dangerous, exploitative nature of the oil industry, in the romance fiction by and for women the intersections between gender, oil, and the text appear to be far more taken for granted. Romance as a genre deliberately showcases the pleasures of the text, including the private female pleasure of reading alone, the promise of love and marriage, and the staging of erotic fantasies. Time and again, these fantasies feature elements of petromodernity, from sex scenes in the back seats of cars to flirtations on the side of the road. In what follows, as I move through a series of romantic motifs that emphasize the pleasures of and in the texts, those points of intersection become visible when the text puts them on display or when we're attentive to them in ways the text might not anticipate. This approach, in revealing the petroculture that intimately structures our daily lives and loves, lets us see romance as one refinement of our petro-imaginary and see petrofeminism as a necessary development in theories of petrofiction that drastically underrepresent women as consumers, producers, and reproducers of oil and petroculture.

What Women Want

When Janice Radway's now-classic *Reading the Romance* was first published in 1984, it gave theoretical form and expression to a genre and its readership that had been ignored by academia. Radway was interested in "what women want from romance fiction" and how they are able to "circumvent the industry's still inexact understanding" of that.[5] In the course of conducting interviews with her informants, Radway came to realize that the "act of romance reading" mattered more to them than the content they were reading; what they wanted was escape from the day-to-day (a temporal condition characterized by a lack of privacy, sparse free time, and frequent interruptions).[6] Radway describes the books as "eagerly read . . . by women who find quiet moments to read in days devoted almost wholly to the care of others."[7] Her women read for the same compensatory reasons—and to escape similar circumstantial limitations—that the Victorian woman read *Pride and Prejudice* or *Jane Eyre*. *Jane Eyre*, in fact, opens with a scene in which the private female pleasure of reading alone is violently interrupted by a male character. While we certainly don't want to collapse the experiences of white nineteenth-century Englishwomen, white twentieth-century Midwesterners, and Black twenty-first-century Nigerians, their reading practices are nevertheless on a continuum, even allowing for radical differences in location, literacy, and leisure.

Studies in petroculture have enjoyed an academic vogue in recent years partly because the topic allows for connections across and between genres and disciplines. But gendered relations to oil have only just begun to be studied—Heather Turcotte's work on petrosexual politics argues that gender violence is a necessary precondition for petro-violence, and Sheena Wilson looks at how images and concepts of women are "systemically co-opted to serve national and international petro-politics."[8] What I'm calling *petrofeminism*[9] may be more immediately visible in the solidarities and alliances forged through and modeled by more explicitly political fictions about oil,[10] but romance fiction, despite its many retrograde qualities, may also be a site of feminist solidarity, whether through its readers who share an interpretive community or through its writers, as in the case of the Kano group in Northern Nigeria, who write in the shared space of the cooperative and find self-determination in adult literacy and professional success. This particular strand of feminism—shaped by, reactive to, and corrective of a petroarchy—suggests that romance and pleasure are as much oil relations as are dirt, violence, and degradation. Petrofeminism highlights the constructive potentials of writing, love, and care in the service of various kinds of liberation ranging from the individual (pertaining to selfhood and sexual identity) to the epochal (pertaining to anthropocenic concerns like mobility and energy independence). Because petrofeminist theory and fiction are both site-specific, we must inevitably speak of petrofeminisms, plural. In Nigeria, concerns about the importation and imposition of white Western feminism have most recently surfaced regarding Adichie and her work,[11] and the African Islamic feminism of Northern Nigeria, as Shirin Edwin calls it,[12] will be different again from feminist expressions in the South. Nevertheless, petrofeminism, while being a retort to globalized corporate self-interest, might also turn out to be a homegrown feminism appearing in homegrown writing.

A search for "Nigeria" on the website for the romance imprint Mills & Boon returns two hits, both for books starring Mack Bolan, a heavily serialized character who fights terrorism the world over. In *Insurrection*, Bolan must "smash al Qaeda's hopes of building yet another major African power base,"[13] and the blurb for *Conflict Zone* tells readers that "Nigeria is rich in oil, drugs and blood rivals—on both the domestic and international fronts. Mack Bolan's ticket into the chaos is a rescue operation involving the kidnapped daughter of an American petroleum executive."[14] We know that Nigerian women are reading Mills & Boon books:[15] Adichie reports that "every girl who grew up in Nsukka when I was growing up read Mills & Boon. I think maybe I read 200."[16] But they are not finding themselves represented in them;

even the Mack Bolan series is not strictly romance but action-adventure published by another division of Harlequin Books. There appear to be no Mills & Boon romance novels set in Nigeria. Thus in this essay I read a spectrum of Nigerian romances, from the self-published *A Heart to Mend* (2009) by Myne Whitman[17] to *Americanah* (2013) by Chimamanda Ngozi Adichie—a book that, through acclaim, has risen above the stigma of romance novel into the stratum of literature—to *Sin Is a Puppy That Follows You Home* (1990) by Balaraba Ramat Yakubu, the first Hausa novel of any sort translated into English.

In *Americanah*, oil fills in background details—a given—constituting part of the market economy in the sections of the book that take place in Nigeria and expressed in the hope that an oil company would rent a block of flats ("Don't worry. . . . God will bring Shell")[18] and in dialogue overheard at a party ("But you know that as we speak, oil is flowing through illegal pipes and they sell it in bottles in Cotonou! Yes! Yes!").[19] In her essay on the novel, Katherine Hallemeier remarks on how it marginalizes global economic history in favor of the central love story: "By treating the political economy as a minor plot point in the romance between Ifemelu and Obinze, Adichie's novel belies expectations that African literature ought to do otherwise."[20] The book, while not marketed as a romance, has many of the classic hallmarks of one[21] and has been called one in numerous reviews.[22] Adichie herself has both claimed and rejected that label, saying, "This is in the grand tradition of Mills & Boon but also it's the anti–Mills & Boon."[23] When Ifemelu and Obinze meet at a dance party, Adichie makes a similar linguistic move, putting the idea of romance under erasure by writing, "Ifemelu thought Mills and Boon romances were silly, she and her friends sometimes enacted their stories." Despite Ifemelu and her friends thinking romance novels are silly, they reenact them anyway. The run-on sentence in the preceding quotation strikes me as purposeful; these things coexist in the same breath without contradicting each other. By making direct, if ironic, reference to the genre, the author flirts with it:

> Ifemelu thought Mills and Boon romances were silly, she and her friends sometimes enacted their stories. Ifemelu or Ranyinudo would play the man and Ginika and Priye would play the woman—the man would grab the woman, the woman would fight weakly, then collapse against him with shrill moans—and they would all burst out laughing. But in the filling-up dance floors of Kayode's party, she was jolted by a small truth in those romances. It was indeed true that because of a

> male, your stomach could tighten up and refuse to unknot itself, your body's joints could unhinge, your limbs fail to move to music, and all effortless things suddenly become leaden.[24]

Both character and author appropriate the romance novel. For the character, romance novels permit safe experimentation with gender identity and budding sexuality; for the author, they allow a self-aware co-optation of the genre's discourse: "'This is really corny but I am so full of you, it's like I'm *breathing* you, you know?' he had said, and she thought that the romance novelists were wrong and it was men, not women, who were the true romantics."[25] We see a similar moment in Whitman's book, suggesting self-conscious romance as a postmodern genre, registering itself and reflecting on the act of reading.[26] The postmodern romance nudges the pleasure of the text from the readerly toward the writerly. After dancing, Ifemelu next tells Obinze that what he said sounds like "the kind of thing you read in a book,"[27] and Yogita Goyal notes that "questions about reading and reception are themselves staged in *Americanah*, which embeds an ongoing critique about books, how they're read, and what they do or fail to do in the world."[28] She goes on to say of Ifemelu: "Each of her romances . . . is mediated by a set of reading protocols."[29] With its interspersed blog posts (extended beyond the bounds of the traditional novel form as an actual blog embedded in Adichie's website), the book upends any usual protocols, in keeping with what—and how—women really want to read now.

Being Moved

Perhaps Nigerian women are reading Ifemelu's blog on their phones. In 2014, UNESCO published a report called *Reading in the Mobile Era: A Study of Mobile Reading in Developing Countries* that describes how the globally ubiquitous mobile phone gets used as a reading device and could be further leveraged as a delivery system for books. The report quotes another report by the Global System for Mobile Communications (GSMA), "a trade body that represents the interests of mobile operators worldwide":

> The study shows that mobile reading represents a promising, if still underutilized, pathway to text. It is not hyperbole to suggest that if every person on the planet

> understood that his or her mobile phone could be transformed—easily and cheaply—into a library brimming with books, access to text would cease to be such a daunting hurdle to literacy. An estimated 6.9 billion mobile subscriptions would provide a direct pipeline to digital books.[30]

While "pipeline" may be simply a metaphor, it may also not be, especially considering that the report quotes the GSMA frequently throughout. It is undeniably in the GSMA's best interest that more phones are used and that more uses are found for phones. The cell phone is not only a commodity but also a petroleum product, and a report like this inadvertently (or not) promotes the exploitation of an emerging African marketplace for a number of commodities in conjunction—the book, the mobile phone, and oil.[31]

What Radway calls the "event of reading"[32] is always contextual, and in the case of Nigerian romance readers, oil literally greases the engine of the text's conveyance, since Nigerians read while mobile in more than one sense. They read on their mobile phones, and they read on the move, while commuting. Mobile phone penetration in Nigeria in 2019 is estimated at 87 percent of the population, according to Jumia, a Nigerian e-commerce website, and the country has one of the highest mobile and online readerships in the world, with those numbers only increasing (by 32 percent from 2017 to 2018 alone).[33] Although the UNESCO report has therefore been outpaced statistically, anecdotally it is still germane. The report invites us to "Meet Nancy,"

> a mobile reader in Abia State, Nigeria. Nancy is 20 years old and loves to read. Nancy's favourite book is *A Heart to Mend* by Nigerian romance author Myne Whitman. Nancy began reading on Worldreader Mobile in May 2013, and that month she spent 10 hours reading. In June, Nancy read on Worldreader Mobile for over 40 hours. When asked "Do you think that you read more now that you can read on your mobile?" Nancy replied, "I do not think that I read more—I know that I read more."[34]

The dual technologies of the mobile phone and the car conspire to associate reading and petromodernity in new and possibly unforeseen ways that register in the literature both blatantly and latently, and in event and content. Wendy Griswold, author of *Bearing Witness: Readers, Writers, and the Novel in Nigeria*, for instance, informs us that the single most common scene in the Nigerian novel is the traffic jam. Indeed, in Whitman's book (freely available to download or read online) much of the connective tissue of the plot involves anticipating traffic, negotiating traffic, or sitting in traffic. Our protagonist, Gladys, arrives in Lagos on a "rickety yellow bus" that takes an hour to cross the bridge into the city, affording a view from its "smudged windows" of a polluted petroscape that will form the backdrop to the story.[35]

Nigerian traffic has taken on mythic proportions, described as an "apocalyptic scene" in media accounts such as the *The Atlantic*'s "World's Worst Traffic Jam" article.[36] Such language is in keeping with a trend to depict African realities as dystopian sci-fi futures that have already arrived. In what is not just a car culture but a culture of "chronic gridlock,"[37] the portability of the mobile phone fits. As though to illustrate this, in *Americanah*'s frame story, when Ifemulu tells Obinze she is moving back to Nigeria from America (thus becoming an Americanah), he reads her email while stuck in Lagos traffic. As an Americanah, one who "look[s] at things with American eyes,"[38] Ifemelu feels "assaulted" by the city upon her return, including by "the yellow buses full of squashed limbs" and the pervasiveness of mobile phones: "When she left home, only the wealthy had cell phones, all the numbers started with 090, and girls wanted to date 090 men. Now, her hair braider had a cell phone, the plantain seller tending a blackened grill had a cell phone."[39] This is proof that cell phones are everywhere and, more pertinently, that they facilitate romance.

The UNESCO report is particularly interesting in its description of who reads what on their phones. Clearly, readers want to be emotionally moved while physically moving, since romance tops the list, with phrases relating to romance (e.g., "sex," "love," and "Romeo and Juliet") among the most popular search terms and romance novels composing nineteen of the top forty books read.[40] If genre is a set of rules for the production of meaning,[41] then the rules of romance novels show up across the genre, no matter whether they are from the Harlequin "fantasy factory"[42] or homegrown Nigerian ones (Ifedigbo).[43] Pamela Regis, in *A Natural History of the Romance Novel,* outlines eight essential elements all romance or courtship novels share. But although the conventions are persistent, they are not static, so that, for instance, even *Jane Eyre* "demonstrates the flexibility of the form."[44] Ann Rosalind Jones has studied how Mills & Boon novels absorb and accommodate feminism. Jones surmises that the romance format is particularly rigid when it comes to the conventions through which the hero is constructed: he is still older, richer, wiser in the ways of the world, and more sexually experienced than the heroine. In Nigerian fiction, these rules hold, but the "urban petroscape"[45] modifies them. Scenes that associate the eligible, desirable hero with his car drive the narrative, and while the narrative perspective still privileges the male gaze,[46] he now habitually looks at the woman on the side of the road from inside his luxury vehicle.

In *A Heart to Mend*, our heroine Gladys first notices her hero when his car pulls out of a garage:

> A black Mercedes S class model with tinted windows purred out. She stared at it in appreciation as it turned towards her, distracted from her quandary for the moment. She took a step back, startled, when the car stopped beside her. Her mouth almost gaped when the back passenger window rolled down. She couldn't help but note the looks of the man who stared back at her. He was very attractive, probably in his mid-thirties, with a certain stamp of authority all over him. His hair was bushier than she usually saw on men his age, but it was full and well-combed. The line between his deep set golden-brown eyes looked ingrained over the strong nose and pink lips accentuated by his dark skin.[47]

Both car and man are made objects of desire by Gladys's appreciative gaze, a gaze reciprocated by him staring back at her. But although they exchange looks, it's not an even trade, since he then scrutinizes her before offering her a lift,[48] retaining his patriarchal powers of selection and control. She can find him attractive, but it's his scrutiny of her that decides their fate. Later, when he drives by and she wonders if he'd seen her, we're told that "Edward had seen Gladys alright. How couldn't he have done so when he'd deliberately altered his route to pass through this street again?"[49]

When she gets into his car, having necessarily demurred first, she subordinates herself by making herself look away from him: "Gladys forced her gaze away and glanced around. The large sports car was spacious and lavishly appointed. She admired the shiny fixtures and automatic controls between the front bucket seats. The whole interior was fitted with smooth tan leather. She could imagine her car-loving brother's envy when she told him about it later."[50] When she turns her attention to the car's luxurious interior, her description lingers over the vehicle in the same way it lingered over the male body. A prosthetic extension of the self, the car signals class, status, and success, all necessary matchmaking qualifiers. It's noteworthy how often cars crop up when Jan Cohn lists the property inventories of romance fiction's desirable bachelors in her study *Romance and the Erotics of Property*.[51] When Alhaji Abdu gets his comeuppance and loses everything in *Sin Is a Puppy That*

Follows You Home, foremost on his list of losses is his automobile: "He had no car; no stall; no merchandise; no money."[52] And, in keeping with *Americanah*'s deconstructionist approach to romance, the elusive love interest in that novel confesses early on that his acquisitions have begun to make him "feel bloated . . . , the family, houses, the cars, the bank accounts."[53] We see both Obinze in *Americanah* and Edward in *A Heart to Mend* being driven by hired drivers, and for Edward his ability to afford this is part of what makes him desirable. A hero straight out of the Victorian mold, his name says it all—Edward Bestman is the best man for Gladys. He shares a first name with *Jane Eyre*'s Edward Rochester and a history as an orphaned child who makes good in the world with *Wuthering Heights*' Heathcliff. He holds out the promise of elevating Gladys's position through marriage in that same tradition—something the novel winks at in a moment of metanarrative: "*Dreamer*, her inner voice mocked, *you've read too many romance novels. This guy drives the latest car models and lives in the posh part of town. He's not going to notice you.*"[54]

The automobile accelerates the erotics of the text, including a latent homoerotics that the mainstream romance can't otherwise come to terms with, through the triangulation of desire as Gladys imagines her "car-loving brother's envy." In imagining this, she calibrates her own lust, as we learn that "her brother had gone crazy when she described that car. But she didn't tell him about the handsome owner who was more remarkable than his car."[55] When the hero recalls their encounter, part of her appeal is that she admired his car; it's completely wrapped up with how she looks (both in the sense of her physical appearance and in the sense of her gaze): "The way her nose turned up at the tip; her delicate eyelids over large dark eyes; the wonder in her gaze as she had admired the car; the way her skirt had ridden up just so to reveal smooth, round knees, and give a hint of other shapely curves too. His heart rate hiked up in a pulsing beat of urgency—a marker of his physical attraction."[56] Lending new meaning to the term *autoerotic*, the book stages love scenes in the car:

> She met him halfway as he bent his head to kiss her again. Her lips parted in welcome, even as he moved his hand around her ears to her collarbone and then lower. He caressed the top of her heaving breasts for a while and then slid his fingers beneath the bodice of the top to touch her breast. Her chest rose and fell as her breath came faster and when she gripped his biceps, his lower body came fully to life. He just had to feel those breasts against him and as he pressed her back into the bucket seat, she moaned against his lips.[57]

The car's intimate space provides a literal vehicle in which the clichés of the genre can unfold.

Love

A world away from Lagos but still in Nigeria, the Northern city of Kano is epicenter to the phenomenon of *littattafan soyayya*, or "love literature." Although these novels function within the constraints of a strict Islamic culture and therefore diverge in all sorts of ways from Mills & Boon–style romance novels, they nevertheless still share a conventional preoccupation with love and marriage. In *Privately Empowered*, her discussion of Islamic feminism in Northern Nigerian fiction, Shirin Edwin takes Novian Whitsitt's point that the novels mark a confrontation with dramatic social change while emphasizing that "the personal and private empoweringly serve as the motor for an entire socio-literary apparatus," rather than the public and the political.[58]

Whitsitt calling the books "vehicles" for the writers' social concerns, and Edwin talking about the "motor" for them being "quotidian private preoccupations"[59] remind us that these books also occupy a petroscape in which cars and the men who drive them feature prominently. This is amply illustrated in *Diagram of the Heart*, Glenna Gordon's photo-essay about the Kano authors, by an image of the cover of the love book *Mardiyya* that features the author's name twice: "Zuwaira Dauda Kolo (Mrs Bashir Ishaq Zugaci)," presumably for decorum. Above this image is a picture of a silver sedan alone against a green backdrop.[60] Gordon's photograph simply shows two side-by-side copies of the book on a lace tablecloth, and the doubling of the image doubles the import of the featured cars. *Littattafan soyayya*, a subgenre of Kano market literature (named for the marketplace where it is sold), has garnered some critical attention, but not much, if anything, has yet been said about the role oil plays in it. This is almost certainly because Nigeria's oil industry has to date operated solely in the South, although it's been reported recently that new Northern wells show prospects of crude.[61] The case for petrofeminism in the oil-rich (or oil-dependent, depending on your perspective) South of the country is more self-evident than in the oil-poor and impoverished North, but the literature makes the case for itself—it is preoccupied with the automobile *despite* not being set in a place saturated by oil rather than because of it.

Sin Is a Puppy That Follows You Home by Balaraba Ramat Yakubu, originally published in 1990, was the first Hausa novel translated into English, in 2012. Concerned primarily with the female characters' matrimonial statuses, it gives us a number of related story lines, including that of a young woman, Saudatu, who aspires to be married before she finishes school. "I don't trust men who like to give girls lifts in their cars," she says, and she is told, "You're right—you should be wary of the ones that want to give you a ride—they're all bastards. They just enjoy corrupting girls. Take a taxi, or walk, but never get into someone's car alone."[62] Soon thereafter, in a scene reminiscent of *A Heart to Mend*, "an Alhaji drove by in a flashy gold-coloured car. They locked eyes for a moment, but she looked away, he passed her without slowing down. She saw him looking back at her, though, as he drove away."[63] Their exchange of glances preserves her modesty—which is reinforced when they meet again and he adjusts his rearview mirror to steal looks at her without her knowledge or reciprocation—but suggests the start of sexual tension between them.[64] In keeping with a context in which everything is commodified, including the women in the story who are constantly called "worthless," her interest in him is based entirely on his car, which "looked expensive and well-maintained."[65] This exaggerates Cohn's idea that the "heroine must not only be aware of the hero's display of consumer goods, she must see the hero in part *through* that display."[66] Luckily for her, he is both rich and kind (unlike some of the other men in the book, who really are bastards), and she gets to marry up out of poverty as his second wife.

In Edwin's discussion of *The Virtuous Woman* by Zaynab Alkali, she describes a similar caution given to Nana, the protagonist of that book, who is told, "'Remember all the things I have been telling you about long journeys.' . . . 'What did I say about accepting favours from strangers?' . . . 'That includes free car rides, monetary gifts and clothes.'"[67] Edwin reads the book, a deliberately moralistic novel intended for adolescents,[68] for its African Islamic feminism that comes to the fore especially when Nana declines a lift from a stranger. Edwin enumerates the qualities that "comprise strong feminism," including Nana's courage, strength of character, decisiveness, and dignity, all of which happen to be made visible by "her steadfast refusal of a free ride."[69] She makes this choice on day one of an empowering journey toward school and a scholarship, a journey not just feminist but petrofeminist in its enabling mobility.

When Charlotte Brontë and her siblings were young, they wrote stories of imagined worlds in miniscule hand in tiny books. The spatial constraints they imposed on their juvenilia reflected their limitations as children and as girls in a society that literally made no room for women writers. The circumstances of their literary production seem not unlike that of the women writers of Kano, who, according to Gordon's introduction to *Diagram of the Heart*, are not supposed to leave their houses.[70] If the comparison to cloistered Englishwomen sounds like overreach, it's worth pointing out that one of the authors Gordon features says she "loves Jane Austen novels"[71] and that another has been profiled by *Time* magazine as "Kano's Jane Austen."[72] Spatially constrained, all these women let their imaginations loose.

Despite laments that the book in Nigeria is an "endangered artifact,"[73] the technology of the book holds its own against the digital technology of the mobile phone, which we can see in the final image in Gordon's book of a Hausa woman reading a romance novel on the train from Lagos to Kano. In transit, she reads a hard copy of the book by the light of her mobile phone, harnessing the newer technology in service to the old.[74] Like the Brontës, the Kano writers capitalize on the book form as a scalable, portable object, often handwriting their texts in small composition books, transcribing them via computer, and self-publishing them. In her essay "Reading Romance Novels in Postcolonial India," Jyoti Puri affords us a cross-cultural comparison[75] when she describes young Indian women's negotiated reading habits, which include ignoring remonstrations from parents, delaying reading the material, and concealing their consumption. One such young woman, "Reshma, who loves 'intense, meaningful, hot, sexy stories,' says of her parents, 'They don't know.' She reads the novels in the bathroom, in the train, and in college."[76] Puri contemplates whether the act of reading romance novels can be considered cultural resistance or a challenge to the hegemonic order but concludes that "the act of reading is limited as a political strategy" since it "remains an isolated, individualized activity" and "may contain and neutralize women's discomfort with their realities."[77] The event of reading matters here, however. To read in liminal spaces, like on the train, is to practice and perhaps even to model behaviors not sanctioned by mainstream authorities, who, in the case of the Northern Nigerian authors, include the morality police and the Ministry of Education as well as parents and spouses. In addition, the Kano women's writers' cooperative confounds the assumption that reading and writing are isolated activities.

In Jones's essay on Mills & Boon, she argues that although the demands and debates associated with feminism produce striking ambivalence in these novels, she does see real innovation over time, even if those changes ultimately contain feminism's radical potential and don't negate the basic premise of the genre, that is, that the greatest goal and pleasure in a woman's life is the love of a good man.[78] One of these changes is a heroine's commitment to her work, and Jones gives the example of a Harlequin romance called *Maelstrom*, in which both hero and heroine are petroleum engineers who risk life and limb to cap an exploding well. Novian Whitsitt's study of love literature lets us trace in it a parallel trend "within a working paradigm of African feminism,"[79] specifically an Islamic Hausa breed of feminism that promotes the working woman who can "fall in love and lead a blessed life of motherhood, career, and opulence beyond her expectations."[80]

Dirt, like work, is traditionally gendered, so that if women's dirt is household dust, the dirt of oil is masculinized. Thinking of petroleum through feminism, however, lets us see the dirty work of oil as being as much women's as men's. This is true not only at the level at which Nigerian women siphon oil from the pipeline to sell by the side of the road but also in the highest ranks of the industry. Although the Nigerian oil industry is still a "boys' club,"[81] the number of female oil executives in Nigeria continues to grow, part of an indigenization plan for the industry. Figures such as Folorunsho Alakija tend to make the news. According to *Forbes*,

> Alakija is worth a staggering $1.73 billion . . . [,] making her the fourth richest person in Nigeria and second richest woman in Africa. She is the Vice Chair of Nigerian oil exploration company, Famfa Oil, which shares a joint partnership agreement with international giants Chevron and Petrobras. With a 60 per cent stake of block OML 127 of the Agbami field, one of Nigeria's largest deepwater discoveries, Famfa Oil produces approximately 250,000 barrels of crude per day.[82]

It's no coincidence that in *A Heart to Mend*, Gladys works hard to qualify for a position with Zenon Oil, a choice reflecting the country's total immersion in the oil industry, a bigger profile for women in it, and the romance novel's increasing interest in women's work.

Corruption and scandal run rampant in Nigerian society, with some of the newly high-profile women in oil ensnared in this other sort of dirty work, such as former petroleum minister Diezani Allison-Madueke. She was charged with money laundering in a case involving the misappropriation of funds from the Nigerian national oil company NNPC in a bribery scandal to keep Goodluck Jonathan in office. In *Slow Violence*, Rob Nixon talks about "the ongoing romance between unanswerable corporations and unspeakable regimes,"[83] which this corruption amply illustrates. Nixon's choice of the word "romance" implies an unsavory relationship but also speaks to a secondary sense of the word—that is, "an extravagant fabrication; a wild falsehood, a fantasy."[84] Any work of romance fiction is a fantasy, remote from everyday life,[85] but this meaning rings especially true when the promise of sentimental love is premised on a romantic scam.

Slickness

The subject of two *Forbes Africa* profiles, Alakija is described as a "slick oil baroness" on one of the magazine's covers,[86] with the punny modifier "slick" introducing a query the accompanying article does not ask—to what extent is she slick because she's a smooth operator and to what extent is she slick because she's a swindler, having greased palms along the way and benefited from corruption at the highest levels? Regardless, slickness, with all its connotations, works. The latest iteration of Nigeria's infamous scamming economy is the romance scam, the new 419 (so called after Article 419 of the country's criminal code, to do with fraud),[87] in which scammers frequently pose as oil rig workers, since it's a profession that covers a multitude of sins, including absenteeism, lack of communication, and the need for funds. *Wired* magazine calls it a "macho cover story that allows them to fade in and out of victims' lives at will."[88] A microcosm of the global scam that is Big Oil, these smaller scams represent the latest installment in the story of leveraging oil for profit at the expense of the vulnerable—in this case, those most susceptible to romance. Cyberpsychologists say that women who fall for online romance scams "tend to score highly on tests that measure how much they idealize romantic love."[89] A set of guidelines available on Facebook on how to avoid being scammed uses the extended metaphor of the "road to romance," which might be "slippery" or "hazardous."[90] But the scammer's slipperiness or slickness is also his appeal.

Accounts of romance scams read like mass-market paperbacks. "Mike Benson" was "a dashing oil worker to whom she'd sent around $14,000 over the preceding months," the *Wired* article gushes in pink prose.[91] Another scammer, posing as someone named Duke McGregor, claimed to be a mechanical engineer with Transocean offshore drilling contractors: "When he wasn't working on North Sea oil rigs, he enjoyed reading classic novels, playing with his tiger-striped tabby cat, and strumming a heart-shaped guitar." Duke's handsome profile photo allegedly showed a middle-aged man with a ruddy face, strong black eyebrows, and a welcoming gaze.[92] The victims of these scams, we are told, often share a particular psychological trait: an exceptional faith in the existence and importance of romantic destiny,[93] which is of course also the reason why women read romances.

Drilling Down

Hitchcock's law of oil states that "it must be opaque or otherwise fantastic," but in romance novels these things are simultaneously true: the fictions depend on an enabling, mobilizing, largely invisible petroculture, and when oil is visible, it is fantastically so, as in the rush to cap the gusher in *Maelstrom*.

In "Petrofiction," his 1992 trailblazing essay, Amitav Ghosh observes that "the history of oil is a matter of embarrassment verging on the unspeakable, the pornographic."[94] The pornographic qualities of oil extend to an interest in the money shot. Textual representations of oil gushers such as scenes in Upton Sinclair's 1927 novel *Oil!* or its 2007 cinematic version *There Will Be Blood* have been read as adolescent ejaculation fantasies,[95] wherein the spectacle of the blowout occludes the slow violence[96] of the oil ordinary.[97] Of a similar moment in the romance canon, Jones writes that *Maelstrom*'s "crisis is genuinely exciting: the couple risks life and limb to cap an exploding well, a spectacular accomplishment."[98] The relationship, however, between romance and porn (and therefore, by extension, between romance and oil) is highly contested. In 1979, Ann Barr Snitow famously equated romance novels with porn for women, arguing that because they are formulaic, "the novels have no plot in the usual sense."[99] A study in the *Review of General Psychology* makes virtually the opposite argument:

> Unlike romance novels, pornography does not really have a "plot." Instead, they typically contain a loosely connected series of sex scenes, each of which usually ends with a visible ejaculation, the "money shot." As a result, pornography has as many

climaxes as it does scenes (at least). A romance novel has only one climax, the moment when the hero and the heroine declare their mutual love for one another.[100]

Another study, entitled "'She Exploded into a Million Pieces': A Qualitative and Quantitative Analysis of Orgasms in Contemporary Romance Novels," published in the journal *Sexuality and Culture*, has it that "female characters were significantly more likely to be depicted experiencing an orgasm during a sexual encounter than male characters."[101] Perhaps the difference lies in degrees of visibility—the money shot and the geyser are visibly spectacular, while the female orgasm and the everyday backdrop of a petroculture usually are not. Maybe the difference is one of degrees of economic conventionality—the romance novel is, fundamentally, a fetishistic capitalist fantasy that must play out within the monetary mainstream to be fulfilling. Anne Cranny-Francis puts it this way: "Romantic fiction is the most difficult genre to subvert because it encodes the most coherent inflection of the discourses of gender, class and race constitutive of the contemporary social order; it encodes the bourgeois fairy-tale."[102]

That bourgeois fairy tale pertains regardless of setting; every study of romance fiction alludes at some point to its formulaic quality (e.g., "The Eight Essential Elements of the Romance Novel"),[103] which is why the romance novel as genre will never entirely spill over from readerly pleasure to writerly jouissance. The single constant is not the attraction, the barrier, or the declaration but the highly conventional pursuit of financial bliss. Everything else—the launching of a career, the happy ending, the heterosexual marriage, the promise of offspring—sublimates itself when an economic reading is foregrounded. Thus, Katherine Hallemeier reads *Americanah* as about the "pursuit of capital and love alike":[104] "*Americanah* presents an alternative, utopic vision of global power in which the United States stands as a foil to the promising future of late Nigerian capitalism."[105] In this dream of Nigerian capitalism, oil would presumably be part of the success story, but Hallemeier gets through her entire essay about capital in the novel without any mention of it.[106] Oil as generative backdrop is so assumed that neither critics nor characters see it. It does not perform itself; it is not pornographically visible. There is a moment late in the novel when Obinze warns that oil companies aren't doing Nigerians any favors,[107] but his success in real estate is predicated on renting to oil men. Everyone in the room has made a fortune from some aspect of the petro-state: Eze, "the wealthiest man in the room," is "an owner of oil wells,"[108] and "Ahmed had leased strategic rooftops in Lagos just as the mobile phone companies were coming in, and now he sublet the rooftops for their base stations and made what he wryly referred to as the only clean easy money in the country."[109] Obinze's fictional warning about "big oils" planning to move offshore and leave onshore operations to the Chinese is framed by the financial press in real life as an indigenous opportunity.[110] "Homegrown" may seem an inapt descriptor for oil (better suited to, say, agriculture), but it is used with some frequency to describe a new phase in Nigerian oil production, one in which local entrepreneurs take over operations. In the dream economy that Adichie's characters aspire to live in, domestic oil control is the material correlate of the textual development that is the homegrown Nigerian romance novel.

Love in *Americanah* does cross boundaries, breach borders, and challenge norms,[111] but it does so within a conventional economic system and is fundamentally conservative in its interest in "manifesting capitalism."[112] This is even more true of the other romance novels read here, although they may actually have more radical potential, even though they are more traditional. *Sin Is a Puppy That Follows You Home* and other love books, for instance, promote female education and have been credited with a rise in literacy among Hausa women.[113] They are also much more likely to meet with censorship and public criticism, according to Whitsitt. Romance as a genre tends to play it safe, and these feminist fantasies imagine oil futures that create a safety net for women's welfare, well-being, and being well-off. Petrofeminism operates within, not outside of, an oil economy, making space inside it for women to claim what is rightfully theirs without having to siphon it off riskily, a space in which the romance of women's work is a job at the oil company and in which a romantic relationship is not possible until the love interest returns to Nigeria to make his fortune renting properties to Shell. The romance stories of Northern and Southern Nigeria offer glimpses into competing modernities, quite different in some essential ways and yet both products of a petroculture shaped and scribed by women. We can read for romance elsewhere too: in short stories by writers such as Sefi Atta and Nnedi Okorafor that demonstrate love and care for nation, community, and environment. Those short stories sabotage the status quo, but the romance novels deliberately do not, playing within received forms of genre and capital. However we understand the petro-imaginary, it must ultimately make all of these iterations visible. Petrofeminism lets us see the inseparable nature of women's politics from oil.

Notes

1. Chimamanda Ngozi Adichie, "Don't We All Write About Love? When Men Do It, It's a Political Comment. When Women Do It, It's Just a Love Story," interview by Emma Brockes, *Guardian*, March 21, 2014.

2. Peter Hitchcock, "Oil in an American Imaginary," *New Formations* 69 (Summer 2010): 87.

3. Research in this chapter was supported by a PSC-CUNY Research Award.

4. Helen Kapstein, "Crude Fictions: How New Nigerian Short Stories Sabotage Big Oil's Master Narrative," *Postcolonial Text* 11, no. 1 (2016).

5. Janice A. Radway, *Reading the Romance: Women, Patriarchy, and Popular Literature* (Chapel Hill: University of North Carolina Press, 1984), 46.

6. Ibid., 7.

7. Ibid., 46.

8. Sheena Wilson, "Gendering Oil: Tracing Western Petro-sexual Relations," in *Oil Culture*, ed. Daniel Worden and Ross Barrett (Minneapolis: University of Minnesota Press, 2014), 177.

9. To the best of my knowledge, this term is my coinage. Since I first used it in an early version of this essay, presented at the Cultural Studies Association Conference at Georgetown University in May 2017, it has started to enter the discourse, although not always with my meanings.

10. For instance, Sefi Atta's short story "A Union on Independence Day" uses elements like realistic news headlines ("Nigerian Delta Women in Oil Company Stand Off") to deliberate effect.

11. See Damilola Odufuwa's "Nigeria: The Women Who Reject Feminism" for how this debate has played out on social media among Nigerian celebrities and politicians (October 8, 2008, https://www.cnn.com/2018/10/08/africa/nigeria-gender-wars-twitter-feminism/index.html).

12. Shirin Edwin, *Privately Empowered: Expressing Feminism in Islam in Northern Nigerian Fiction* (Evanston: Northwestern University Press, 2016).

13. Don Pendleton, *Insurrection* (London: Mills & Boon, 2015).

14. Don Pendleton, *Conflict Zone* (London: Mills & Boon, 2015).

15. In second place on the list of top ten books read by Worldreader Mobile users in UNESCO's *Reading in the Mobile Era* report is Mills & Boon's *The Price of Royal Duty*, read by 18,364 people from seven countries (including Nigeria) in three months.

16. Zadie Smith and Chimamanda Ngozi Adichie, "Between the Lines: Chimamanda Ngozi Adichie with Zadie Smith," *Schomburg Center for Research in Black Culture*, March 19, 2014, https://livestream.com/schomburgcenter/events/2831224/videos/45613924.

17. Nkem Okotcha's pen name.

18. Chimamanda Ngozi Adichie, *Americanah* (New York: Anchor Books, 2013), 27.

19. Ibid., 37.

20. Katherine Hallemeier, "'To Be from the Country of People Who Gave': National Allegory and the United States of Adichie's *Americanah*," *Studies in the Novel* 47, no. 2 (Summer 2015): 236.

21. It is what *Writing a Romance Novel for Dummies* calls a "reunion romance," also known as a second-chance romance, and closely follows Shoshanna Ever's "Secret Formula of Most Romance Novels."

22. Additionally, critics have pointed out the "(surprisingly spare) critical commentaries" on *Americanah* (Jennifer Leetsch, "Love, Limb-Loosener: Encounters in Chimamanda Adichie's *Americanah*," *Journal*

of Popular Romance Studies [April 2017]: 3, http://www.jprstudies.org), with one attributing this directly to its romance form: "The reasons of the lack of academic interest in *Americanah* may lie in the very nature of the novel, the plot being quite linear and 'pop' and based on a romantic love story; this aspect, however, may be considered on the contrary positive, since it makes complex gender and 'racial' issues more accessible and visible to common readers." Valentina Scarsini, "*Americanah* or Various Observations About Gender, Sexuality and Migration: A Study of Chimamanda Ngozi Adichie" (master's thesis, Universita Ca'Foscari Venezia, 2016–17), 11, http://dspace.unive.it/handle/10579/10687.

23. Smith and Adichie, "Between the Lines."

24. Adichie, *Americanah*, 69–70.

25. Ibid., 239.

26. For connections between postmodernism, feminism, and romance, see Diane Elam's *Romancing the Postmodern*, although Elam does not discuss romance novels per se. For Elam, the romance is postmodern because it marks and is marked by excess—"its capacity to appear where least expected"—but this theory works only when one is looking for romance outside of romance novels (London: Routledge, 1993), 12.

27. Adichie, *Americanah*, 70.

28. Yogita Goyal, "Introduction: Africa and the Black Atlantic," *Research in African Literatures* 45, no. 3 (Fall 2014): xiii.

29. Ibid.

30. Mark West and Han Ei Chew, *Reading in the Mobile Era: A Study of Mobile Reading in Developing Countries*, ed. Rebecca Kraut (Paris: UNESCO, 2014), 17.

31. See Jennifer Wenzel's "Petro-Magic-Realism: Toward a Political Ecology of Nigerian Literature" on the connection "between Nigerian literary production and other commodity exports." *Postcolonial Studies* 9, no. 4 (2006): 449.

32. Radway, *Reading the Romance*, 7.

33. Chike Onwuegbuchi, "Nigeria, Others Emerge Highest Mobile Online Reading Population, Says Report," *The Guardian*, September 14, 2018.

34. West and Chew, *Reading in the Mobile Era*, 47.

35. Myne Whitman, *A Heart to Mend* (Bloomington, IN: AuthorHouse, 2009).

36. Joshua Hammer, "World's Worst Traffic Jam," *The Atlantic*, July/August 2012.

37. Ibid.

38. Adichie, 475–76.

39. Ibid.

40. West and Chew, *Reading in the Mobile Era*, 52.

41. Morley, quoted in Radway, *Reading the Romance*, 10.

42. Ann Rosalind Jones, "Mills & Boon Meets Feminism," in *The Progress of Romance: The Politics of Popular Fiction*, ed. Jean Radford (New York: Routledge, 1986), 211.

43. Ifedigbo calls Whitman's novels homegrown, although the Nigerian-born Whitman now lives in diaspora in Seattle.

44. Pamela Regis, *A Natural History of the Romance Novel* (Philadelphia: University of Pennsylvania Press, 2007), 85.

45. Stephanie LeMenager, "The Aesthetics of Petroleum, *After Oil!*," *American Literary History* 24, no. 1 (Spring 2012): 66.

46. Jones, "Mills & Boon Meets Feminism," 214.

47. Whitman, *Heart to Mend*, 6.

48. Ibid., 7.

49. Ibid., 30.

50. Ibid., 7.

51. Jan Cohn, *Romance and the Erotics of Property: Mass-Market Fiction for Women* (Durham: Duke University Press, 1988), 19n7, 46, 154.

52. Balaraba Ramat Yakubu, *Sin Is a Puppy That Follows You Home*, trans. Aliyu Kamal (Chennai, IN: Blaft, 2012), 104.

53. Adichie, *Americanah*, 26.

54. Whitman, 29.

55. Ibid.

56. Ibid., 18.

57. Ibid., 72.

58. Edwin, *Privately Empowered*, 28.

59. Ibid., 165.

60. Glenna Gordon, *Diagram of the Heart* (Brooklyn, NY: Red Hook Editions, 2016), 128–29.

61. Initial reports in 2016 were optimistic ("GOOD NEWS: Oil Wells Discovered in Northern States of Nigeria," *Savid News*, Oct. 15, 2016), followed by doubt ("$3b Down the Drain, Oil in North Remains Elusive," *Guardian* [Nigeria], Nov. 6, 2017), followed most recently by renewed (if qualified) hope ("Excitement in Northern State as 7 Wells 'Confirm' Existence of Crude Oil," *Nigerian Bulletin*, May 25, 2018).

62. Yakubu, *Sin Is a Puppy*, 29.

63. Ibid., 30.

64. Ibid., 53.

65. Ibid., 31.

66. Cohn, *Romance and the Erotics of Property*, 46.

67. Alkali, quoted in Edwin, *Privately Empowered*, 122.

68. Alkali, quoted in Edwin, *Privately Empowered*, 116.

69. Edwin, *Privately Empowered*, 123.

70. Gordon, *Diagram of the Heart*, 15.

71. Ibid., 20.

72. Aryn Baker, "Kano's Jane Austen," *Time*, May 17, 2018, https://time.com/collection-post/5277967/farida-ado-next-generation-leaders/.

73. "Nigeria Is a Bookless Country," *Vanguard*, July 7, 2013, https://www.vanguardngr.com/2013/07/nigeria-is-a-bookless-country-prof-emenanjo/.

74. Ibid., 130–31.

75. The degrees of cultural separation here may be even fewer than one would imagine: Shirin Edwin pointed out to me in conversation that it's no coincidence that *Sin Is a Puppy That Follows You Home* was published by an Indian house—it's the kind of conventional narrative that appeals to a mainstream Indian audience, she says. Moreover, as she explains in *Privately Empowered*, the books themselves are heavily influenced by Bollywood films (26).

76. Jyoti Puri, "Reading Romance Novels in Postcolonial India," *Gender and Society* 11, no. 4 (August 1997): 440.

77. Ibid., 441–42.

78. Jones, "Mills & Boon Meets Feminism," 211.

79. Novian Whitsitt, "Islamic-Hausa Feminism Meets Northern Nigerian Romance: The Cautious Rebellion of Bilkisu Funtuwa," *African Studies Review* 46, no. 1 (April 2003): 142.

80. Ibid., 147.

81. Orin Gordon, "Nigeria's Growing Number of Female Oil Bosses," BBC News, September 11, 2014.

82. "A Head for Business and Fists for a Fight," *Forbes Africa*, August 1, 2016.

83. Rob Nixon, *Slow Violence and the Environmentalism of the Poor* (Cambridge, MA: Harvard University Press, 2011), 105.

84. *OED Online*, s.v. "romance, n. and adj. 1," https://www.oed.com/view/Entry/167065?rskey=EY55Et&result=1&isAdvanced=false#eid.

85. Ibid.

86. "Folorunsho Alakija: Slick Oil Baroness," *Forbes Africa Woman*, December 2013–January 2014.

87. Andrew Apter traces a direct lineage from crises in the oil economy to the "rise of the era of the '419'": "As inflation soared, arbitrary exchange values destabilized the very phenomenology of exchange itself." *The Pan-African Nation: Oil and the Spectacle of Culture in Nigeria* (Chicago: University of Chicago Press, 2005), 277.

88. Brendan Koerner, "Online Dating Made This Woman a Pawn in a Global Crime Plot," *Wired*, October 5, 2015.

89. Ibid.

90. MoneyGram, "How to Stop Scammers in Their Tracks on the Road to Romance," Facebook, January 30, 2013, https://www.facebook.com/notes/diskarteng-moneygrado/how-to-stop-scammers-in-their-tracks-on-the-road-to-romance/569773693052013/.

91. Koerner, "Online Dating."

92. Ibid.

93. Ibid.

94. Amitav Ghosh, "Petrofiction," *New Republic* 206, no. 9 (March 1992): 29.

95. LeMenager, "Aesthetics of Petroleum," 74.

96. Ibid, 76, 77.

97. Oil as ejaculate is only one version of the persistent metaphoricity of oil. The commodity's transitive properties (it becomes everything from fuel to plastic to money) and its ubiquity mean that oil is a slippery term, transmuting into "black gold" and even transubstantiating into "the very blood of the nation and its citizens." Apter, *Pan-African Nation*, 277.

98. Jones, "Mills & Boon Meets Feminism," 211.

99. Ann Barr Snitow, "Mass Market Romance: Pornography for Women Is Different," in *Women and Romance: A Reader*, ed. Susan Ostrov Weisser (New York: NYU Press, 2001), 309.

100. C. Salmon, "The Pop Culture of Sex: An Evolutionary Window on the Worlds of Pornography and Romance," *Review of General Psychology* 16, no. 2 (2012): 155.

101. Christine Cabrera and Dana A. Menard, "'She Exploded into a Million Pieces': A Qualitative and Quantitative Analysis of Orgasms in Contemporary Romance Novels," *Sexuality and Culture* 17 (2012): 193–212.

102. Anne Cranny-Francis, *Feminist Fiction: Feminist Uses of Generic Fiction* (New York: St. Martin's Press, 1990), 192.

103. Regis, *A Natural History of the Romance Novel.*

104. Hallemeier, "To Be from the Country," 242.

105. Ibid., 232.

106. Ibid., 237.

107. Adichie, 580.

108. Ibid., 577.

109. Ibid., 576.

110. See, for example, Jeff Grey, "Meet the New Face of Nigeria's Oil Industry" *Globe and Mail*, August 10, 2014.

111. Compare Leetsch, "Love, Limb-Loosener."

112. Hallemeier, "To Be from the Country," 235.

113. Rakesh Khanna, "A Note from the Publishers," in Yakubu, *Sin Is a Puppy*, vi. See also Edwin, *Privately Empowered*, on this.

5.

"We Are Pipeline People"
Nnedi Okorafor's Ecocritical Speculations

Wendy W. Walters

Petroleum extraction, transportation, and consumption follow global routes familiar to the historiography of the African diaspora. Pipelines in the Niger Delta transport crude oil to tankers owned by US-based multinational oil corporations such as Shell, Chevron, and ExxonMobil, delivering this global commodity across trans-Atlantic routes to North American entry points, routes familiar to the history of the slave trade. Whereas European and American projects of empire extracted the lives and bodies of African people in the centuries preceding oil, since the discovery of crude oil in the Niger Delta in the 1950s a different form of extraction has been occurring, and human and ecological futurity continue to be toxically threatened.[1] Global capitalism and its ongoing demand for ever more oil render life in the Niger Delta precarious. Exploitative conditions and ecological devastation are repeatedly resisted by local citizen activists in the Niger Delta as well as by writers of diasporic ecocritical literature. Critical Afrofuturism, or Black speculative fiction focusing on social change, demonstrates the power of narrative to both reveal and warn of ecological risk and its consequences. Two short stories by Nnedi Okorafor, "Spider the Artist" and "The Popular Mechanic," are set in rural Niger Delta communities and demonstrate Black Atlantic cultures of resistance to oil's local and global devastations.

Born to Nigerian parents and raised in the United States, Okorafor describes her creative muse as located in Nigeria; much of her fiction is set in Nigeria and traces out connections between Nigeria and the United States. As she explains, "Being raised as a Nigerian American is all over my work. That hybridity, the conflicts, the similarities—the fusion of those two cultures combining and conflicting—that is why I am who I am and why I write what I write."[2] Reading through the lens of critical Afrofuturism, this chapter examines how Okorafor's short fiction offers a global accounting, linking the effects of oil extraction in the Niger Delta to US consumption. Critical theorists, artists, and writers of Afrofuturism and/or Black speculative fiction describe their art as counterhegemonic and involved in deconstructing Eurocentric imperial projects built on structural racism and social injustice.[3] Lisa Yaszek sees Afrofuturism as a way to "say no to those bad futures that justify social, political, and economic discrimination. In doing so this mode of aesthetic expression also enables us to say yes to the possibility of new and better futures and thus to take back the global cultural imaginary today."[4] Okorafor is a multiple-award-winning author of at least eleven books, including novels, series, comics, and children's and young adult books. Her 2016 novel, *Lagoon*, was a British Science Fiction Association Award finalist for best novel, and *Who Fears Death* (2010) won a World Fantasy Award for best novel. The first book in her Binti trilogy (2015) won both the Hugo and the Nebula Awards for best novella. *Zarah the Windseeker* (2005) won the Wole Soyinka Prize for African Literature. Okorafor rejects the reductive qualities of literary labels, stating, "I'm known for writing speculative fiction novels. This includes science fiction, fantasy and magical realism. These are just labels created by others. Labels I sometimes feel as really reductive. Is my aliens-come-to-Lagos novel *Lagoon* just science fiction? I think it can also be categorized as African Literature."[5] She describes her work as Africanfuturism, not Afrofuturism.

I focus here on two of Okorafor's short stories that feature the oil pipeline in its local and networked materiality. The pipelines in these stories dominate the residential backyards of the characters, linking these local spaces to the global distribution of oil. Envisioning and imagining alternative futures, Afrofuturist (and Africanfuturist) literature and art often turn toward radical ecological critique. Although Okorafor depicts the ecological devastation wrought by multinational oil companies, her stories also deploy an imaginary aesthetic and feature characters who display creative and resistant futurity. Okorafor's use of the speculative enables a critical view of global oil's predations as well as irruptions of African-centered feminist agency and futurity.

Eme, the narrator of "Spider the Artist," sets the scene within petroleum's local infrastructure: "My village was an oil village, as was the village where I grew up. My mother lived in a similar village before she was married, as did her mother. We are Pipeline People."[6] The story here names a multigenerational politics of the pipeline, an epistemology and a positionality that is not new but in fact passed down over time and across villages. This is not just a social knowledge but also an embodied knowledge—it is a biological and genetic knowledge, lived and expressed through oil's effects on the body. I read Okorafor's work as critical Afrofuturism, since it operates "from a standpoint that intersects theories of time and space, technology, class, race, gender, and sexuality and delineates a general economy of racialization in relation to forces of production and apocalyptic, dystopian, and utopian futures."[7] Kodwo Eshun emphasizes the importance of revising global capital's speculations about Africa: "If global scenarios are descriptions that are primarily concerned with making futures safe for the market, then Afrofuturism's first priority is to recognize that Africa increasingly exists as the object of futurist projection. African social reality is overdetermined by intimidating global scenarios, doomsday economic projections, weather predictions, medical reports on AIDS, and life-expectancy forecasts, all of which predict decades of immiseration."[8] Eshun defines these NGO- and MNC-authored scenarios as ideologically demoralizing menaces of a predatory future in which "Africa is always the zone of absolute dystopia."[9] In what follows I aim to show the ways that Nnedi Okorafor's speculative fiction resists depicting absolute dystopia and yet marshals a complex critique of multinational oil corporations operating in the Niger Delta. Attending to speculative scenarios within Okorafor's fiction shows how her work performs the type of chronopolitical intervention Eshun calls on to disrupt the production and distribution of dystopic representations. If Okorafor's fiction may be said to depict a dystopic present, that depiction is always causally linked to American technoscience, and it is always a mobile and shifting present, open to a variety of forms of resistant African-centered futurity.

Scholarship on speculative fiction depicting environmental destruction often engages the terms *apocalypse* and *crisis*. Future-projecting fiction and criticism also frequently take an ecocritical turn, since an apocalyptic future is often the result of human-caused environmental damage. In "Oil Culture," Ross Barrett and Daniel Worden note "the problematic relationships that have taken shape between oil and conceptions of futurity, the deep cultural entanglement of petroleum and apocalypse."[10] But dystopic projections of environmental destruction are only the future in particular locations or for certain subjects—for others they may describe a past and/or an ongoing present. A comparatist framework requires more global timescapes. The transatlantic slave trade can be seen as a historical *apocalypse* for African peoples in many senses of the word. Global capitalism continues to create contemporary conditions of crisis, if not apocalypse, particularly for people living in the Global South and also increasingly in northern geographies such as Detroit, Flint, and North Dakota. A Black Atlantic paradigm stemming from Diaspora Studies, then, provides a guide.[11] In "Environmentalism and Postcolonialism," Rob Nixon stresses the need for "a transnational ethics of place" in ecocriticism, and he cites Black Atlantic studies as a model for this paradigm shift.[12] Nixon develops this global focus at length in *Slow Violence and the Environmentalism of the Poor*, linking "two of Earth's greatest, most vulnerable delta wetlands—the Mississippi and the Niger. In Nigeria's southeastern delta states—which satisfy 11 percent of American oil needs—546 million gallons of oil have spilled to date" during fifty years of drilling. Nixon goes on to note that "Niger Delta communities have suffered the equivalent of an Exxon-Valdez sized spill *annually* for a half century."[13] The oil pipelines so present in "Spider the Artist" and "The Popular Mechanic" contain, transport, and violently eject a toxic fuel close to the lives of Okorafor's characters, fuel that is destined for others across the Atlantic. Recent writing on oil fiction identifies the limits of both nation-state and genre boundaries in representing and analyzing this literature. Petrofiction's "multinational structures, routes, and determinations ensure its contemporary identification as a subgenre of literature more productive under the rubric of 'world literature' than it is under that of any national literary corpus."[14] African diasporic literature also exceeds nation-state boundaries and generates reading strategies attuned to multiple traditions and references, what VèVè A. Clark has called "diaspora literacy."[15] Before oil, literature of the transatlantic slave trade and the plantation complex required a global focus. Readers of Okorafor's fiction encounter a local focus with a global accounting. Her depiction of pipelines in these two stories calls attention to the direction in which the liquid flows and to the limits of its containment.

Oil's relation to dystopia, then, is contextual. In some dystopic projections, running out of oil is the future apocalypse that will change life as we know it. In the Niger Delta, one might argue that the discovery of oil itself created an ongoing apocalypse that changed life as it was known by the thirty-one million people living in what is "the largest wetland in Africa and one of the largest in the world," previously home to a "vast biodiversity . . . and a fragile ecology."[16] Oil multinationals Shell, Chevron, Texaco, ExxonMobil, Total, and Agip(Eni) operate in the Niger Delta, mostly in oil fields that are also the homes of a complex group of ethnic minorities with varying relationships to the Nigerian state.[17] The result is that small and primarily "rural communities of this region host some of the most sophisticated multinationals on earth, and also suffer the direct devastating impact of oil production and oil accidents, making [the Niger Delta] 'one of the world's most severely petroleum impacted ecosytems and one of the five most petroleum-polluted environments in the world.'"[18] Environmental activist and writer Ken Saro-Wiwa's well-known struggle for Ogoni rights arose because his community was "home to six oilfields, half of Nigeria's oil refineries, the country's only fertilizer plant, [and] a large petrochemical plant," a place where life expectancy is just 50 years.[19]

Do these conditions render the Niger Delta already a postapocalyptic space? Reading Okorafor's fiction as critical Africanfuturism not only locates the cause of the environmental harm but further distributes African agency and imagination in a politics of resistance. The Greek etymology of the word *apocalypse* can be traced to "a *revelation* or *disclosure* of the spatial and temporal transformation of the world, not simply a prediction of the 'end times.'"[20] It's also important to consider that "the word has a technical usage that implies a transformation, perhaps in consciousness, by which an existing corrupt socio-ecological order is turned upside down by the astonishing irruption of new hope."[21] This emphasis on the transformative potential of new hope is a key aspect of Afrofuturist art. Uniting the creative and the political, visionary Afrofuturist literature allows a "decolonization of the imagination" necessary to overturn corrupt orders.[22] As Robin D. G. Kelley has documented so well in *Freedom Dreams*, the Black radical imagination has always informed the Black radical tradition, "an unleashing of the mind's most creative capacities, catalyzed by participation in struggles for change."[23] While Okorafor's fiction may be said to "reveal" (to readers in the Global North) the apocalyptic conditions facing those for whom the pipeline has always been visible, in its use of speculative modes her fiction both critiques past and present injustice and imagines futurity and resistance.

"Spider the Artist" and "The Popular Mechanic" locate the petro-violence of the Niger Delta within global circuits of petroleum production and consumption by focusing on the pipeline in all its networked visibility and materiality. Okorafor links her interest in oil pipelines to their global connections: "Oil spills, gas flares, pipeline explosions, poor land management, human rights abuses, the oil companies and the Nigerian government could care less about the land or people. Mind you, Nigeria is the United States' fifth largest oil supplier. So I'm interested in this issue for many reasons."[24] Her speculative fiction set in Nigeria thus stages the oil encounter, and her work requires paying attention to oil and its pipelines from the perspective of those who have been forced to encounter oil for the past half century in visceral, bodily, and unavoidable ways. These two stories center local pipelines, connecting them to the global supply chain of oil consumption. The pipeline is "oil's containing infrastructure," covering the globe with "more than 3.5 million kilometres of pipelines on the planet—more than nine times the distance of the Earth to the moon"—and accounting for nearly 40 percent of the world's CO_2 emissions.[25] Yet Imre Szeman describes pipelines as only recently becoming visible: "Until recently, pipelines have not played a role in politics in large part because they were, on the whole, as socially invisible as they were physically distant and out of sight, neither encountered by the public in daily activities nor featured in their social imaginaries."[26] Okorafor's fiction removes the fiction of pipeline invisibility. For whom are pipelines "out of sight"? Who is the public who does not encounter pipelines in their daily activities? What counts as politics? Indeed Graeme Macdonald notes that the "invisibility" of pipelines "is both culturally 'naturalized' and internationally relative—structurally managed in the interests of an oil-reliant capitalism seeking to extend and perpetuate supply while downplaying the ongoing, exploitative shame of extraction and land dispossession and the inevitable endpoints of burning."[27] Okorafor's literary intervention not only depicts a politics of pipeline sabotage but also literally serves as a pipeline moving the material knowledges of people in the Niger Delta into North America. The "daily activities" and "social imaginaries" of the characters in "Spider the Artist" and "The Popular Mechanic" revolve around the pipelines that run through their neighborhoods. If the pipeline is a metonym for the global web of petromodernity, Okorafor's fiction hacks the pipeline.

In these two short stories the smell of petroleum is acrid, immediate, unavoidable, and toxic: "The smell was overwhelming, *o*! Stung my nose and stuck to my clothes. We all withstood it with runny noses, turning our heads to the side and spitting."[28] "Spider the Artist" takes place in a rural village that has been ravaged by the petroleum industry. Okorafor depicts a dystopic present, representing not only the ecological damage wrought by global capital but also the domestic violence engendered by frustratingly difficult resistance to the petro-state. The narrator links her husband's violence to his alcoholism, caused ultimately by the widespread ecological destruction produced by the presence of the oil companies:

> My husband was a drunk, like too many members of the Niger Delta People's Movement. It was how they all controlled their anger and feelings of helplessness. The fish, shrimps, and crayfish in the creeks were dying. Drinking the water shriveled women's wombs and eventually made men urinate blood.
>
> There was a stream where I had been fetching water. A flow station was built nearby and now the stream was rank and filthy, with an oily film that reflected rainbows. Cassava and yam farms yielded less and less each year. The air left your skin dirty and smelled like something preparing to die. In some places, it was always daytime because of the noisy gas flares.
>
> My village was shit.[29]

Although Okorafor engages in realist representation of the actual environmental degradation documented by many scholars,[30] she writes a speculative narrative of resistance and futurity amid grave threat. There is no absence of life but rather daily interactions with the pipeline infrastructure. The narrator, Eme, describes the tightly bounded domestic space in the story's first sentences, as she is driven out to the backyard by her husband's abuse: "My husband used to beat me. That was how I ended up out there that evening behind our house, just past the bushes, through the tall grass, in front of the pipelines."[31] Okorafor's fiction depicts the many ways these neighborhood pipelines are intimately connected to peoples' wombs, skin, urine, and blood—the narrator explains, "We are Pipeline People."[32] The story is also about ongoing generations whose futurity is imminently threatened. Fishing and farming are made untenable by the presence of petroleum infrastructure; human bodies are made ill; and the night sky is polluted with light.

Okorafor speculates on the logical outcome of petro-capitalism's concern for property over life and profit over people in her creation of a new technological infrastructure developed to police and protect the fallible pipelines: Zombies. The backyard pipeline is patrolled by these Zombies, robotic policers of the oil infrastructure capable of tearing people limb from limb. "They were originally called Anansi Droids 419 but we call them '*oyibo* contraption' and, most often Zombie, the same name we call those 'kill-and-go' soldiers who come in here harrassing us every time something bites their brains."[33] These Zombies are not so much the undead of popular representation as they are programmed robots whose mission is to protect the pipeline, not the people. Like other AI technology, they do not always perform as programmed. In the present of the story, being "Pipeline People" means being caught in the middle of a violent nexus of global capital, protected by neither the state nor the oil companies who extract resources from the land on which they live: "The government came up with the idea to create the Zombies, and Shell, Chevron, and a few other oil companies (who were just as desperate) supplied the money to pay for it all. The Zombies were made to combat pipeline bunkering and terrorism. It makes me laugh. The government and the *oil people* destroyed our land and dug up our oil, then they created robots to keep us from taking it back."[34] Since "oil people" represent the MNCs (the "oil men" of popular discourse), the story shows how the bodies of pipeline people have become the literal conduits for oil people's wealth. The MNCs' profit logics reduce people to pipelines, as their bodies bear the material waste produced by petroleum extraction. The Zombie policers servicing the material value of petroleum's infrastructure protect the pipelines, not the people's bodies.

Eme's village is also a site of gendered violence and exclusion: "No matter my education, as soon as I got married and brought to this damn place, I became like every other woman here, a simple village woman living in the delta region where Zombies kill anyone who touches the pipelines and whose husband knocks her around every so often."[35] In the global web of petromodernity, however, "a simple village woman" has complex situated knowledge and a story to tell. While her story depicts a local effect, Eme extends the critique of petro-violence globally: "Some of these pipelines carry diesel fuel, others carry crude oil. Millions of liters of it a day. Nigeria supplies twenty-five percent of United States oil. And we get virtually nothing in return. Nothing but death by Zombie attack. *We can all tell you stories.*"[36] Eme's citation of these multiple stories points to the power of collective narratives of resistance and truth. Although "Spider the Artist" is set in a small rural village, located by the specific geography of an oil extraction site, this location is intimately connected to the global networks of oil consumption, serviced by the pipelines bringing oil out of the Delta. Cyril Obi and Siri Aas Rustad corroborate the narrator's own knowledge, confirming that "Nigeria is currently the fifth-largest oil supplier to the United States. Two U.S. oil MNCs (ExxonMobil and Chevron Texaco) [currently] operate in the Niger Delta."[37] The history of Western imperialism, colonialism, the trans-Atlantic slave trade, and multinational capitalism link the oil drilled in the Niger Delta to the US petroleum century. Fiction adds one form of narrative supplement to the history of resource extraction, depicting characters with embodied knowledges of the effects of government policies and actions.[38]

Eme offers not only her own but multiple "pipeline people's" stories about the human routes of a global oil supply chain. While the government tells one story about the Zombies' purpose, local people catch the lie and offer their own evidence: "Government officials *said* that Zombies were programmed to do as little harm as possible but . . . I didn't believe this, *na* lie."[39] Okorafor's speculative fiction supplements the local knowledges relayed by Justina Adalikwu-Obisike and Ebere E. Obisike in their study "Communities at Risk: An Aftermath of Global Capitalism." They document multiple examples of similar local knowledges, linked to global actors, in their interviews with Niger Delta residents between 2004 and 2013. For example, one interviewee stated, "Formerly, when you plant crops, they will yield very well but now, the crops do not yield well because of the activities of Shell. A year before last, there was a fire outbreak in our farmland and all the cassava in every farm died. Since Shell came to Nigeria we have been experiencing these problems. They burn down farmlands, pollute the air, water and land and the crops yield is very small. People get sick and they do not have money to go to the hospital and as a result, they die very young.'"[40] Like these residents, Eme knows the truth that's visible behind the concealing language of the petro-state, asserting a local knowledge, inflected with local language, about the real priorities of "oil people."

Petroleum monoeconomies generate forms of resistance ranging from piracy to kidnapping to destruction of oil infrastructure. The response to such resistance is often further violence. Adalikwu-Obisike and Obisike note that "the ecological destruction of creeks and waterways in the remote areas of the Niger Delta is equated to the pollution of the cultural and public sphere by an invasive and extractive petro-state since oil is what gives the nation its foreign exchange."[41] The Nigerian state, backed by the oil multinationals, is notorious in its repression of all forms of resistance. Peaceful protest such as that launched by Ken Saro-Wiwa and the MSOP (Movement for the Survival of the Ogoni People), a movement that gained international recognition, was met with state violence and the murder of Saro-Wiwa and eight others in 1995. Obi and Rustad, among others, define the state as "more of an oil gatekeeper and oil revenue collector, operating in partnership with, and beholden to, oil

MNCs."[42] Despite Saro-Wiwa's and MSOP's connection to a global discourse of rights and environmentalism, the far more powerful workings of global capital with a long history of transnational violence maintain dominance, with the United States playing an active role in suppressing the insurgent militias attempting to claim local minority rights.[43] Eme attributes her husband's violence to the "anger and helplessness" felt by "too many members of the Niger Delta People's Movement."[44] Frustration about the scarcity of fuel for those living amid petroleum's infrastructure leads to other forms of resistance.

"Spider the Artist" and "The Popular Mechanic" are also stories about the potential rewards of bunkering in an oil-scarce economy. When pipelines either burst or were broken, people rushed to collect the toxic liquid, since "fuel in Nigeria was liquid treasure. On the black market it was more valuable than clean water. Thus a burst pipeline attracted desperate pirates of all kinds."[45] The narrator of "The Popular Mechanic" explains that

> Nigeria was one of the world's top oil producers. Yet and still, as the years progressed, the Nigerian government had grown fat with wealth harvested from oil sales to America. The government, to the great detriment of the country, ate most of the oil profits and could care less about what the process of extracting the oil did to the land and its people. On top of all this, ironically, Nigeria's people often suffered from shortages of fuel.[46]

"The Popular Mechanic" is told from the perspective of Anya, a first-year medical student who is home at the end of the semester visiting her mechanic father and schoolteacher mother. The narrative relates a previous story told by Anya's father of how he lost his arm while collecting spilling fuel from a burst pipeline. When someone lit a cigarette, the resulting explosion killed ninety-nine people. Okorafor's speculative turn in this story concerns a cybernetic arm transplant. Echoing historical medical experimentation done by US biomedical and pharmaceutical companies on African diasporic people,[47] "the Americans came offering several million naira to ten one-armed individuals willing to try" the new arms. Of course the new arms came with risk: "The side effects of the transplant were well noted. Periods of delusion, paranoia, and mild incontinence. . . . Anya was sure that none of the results from her father's experience—from the delusions to the incontinence—would reach the American newspapers or scientific journals."[48] Okorafor's story foregrounds her characters' experience and knowledge. As a future doctor, Anya's medical expertise may help prevent future exploitation of Nigerians by US biotechnology/pharmaceutical companies.

"The Popular Mechanic" also represents the materiality of oil pipelines and the intrusion of this volatile infrastructure right under people's noses. When her father goes missing, Anya follows a line of people "all carrying buckets, cups, jerri-cans, large bowls, plastic containers. After a while, she could have also followed her nose, the smell was so strong. It stung and bit at her nostrils and eyes."[49] Anya finds her father, acting as an "Igbo Robinhood," at a spot where the raised pipelines running behind homes have burst. He has used his own mechanical knowledge to make his high-tech American arm into a complex tool capable of breaking open the pipeline. In addition, he's installed a microphone into the arm, by which he is keeping order in the crowd, reminding people to take caution around the highly flammable liquid. The narrator describes a large pool of fuel that is "dark pink, almost like watery blood."[50] The infrastructure of petroleum smells most sharply in this story. "The Popular Mechanic" is replete with descriptions of the materiality and toxicity of oil and its effects on the body: the fuel soaks through clothes; makes skin itch; causes eyes to water, redden, and sting; and makes nostrils burn. When Anya finds her father at the pipeline, he explains to her with clearly incisive reasoning: "Anya, there are conspiracies, stinging wires, implanted computer chips everywhere. But we should still try and do what we can."[51] Although the story links this statement to the delusions he suffers as a medical side effect, he speaks a truth about neoliberal capital. He exacts his revenge, claiming what he sees as payment owed for the side effects produced by the cybernetic arm and telling his daughter, "Those white people owed me three buckets of fuel."[52]

Anya's father generates his own story, instructing his neighbors gathered around the pipeline to "keep your mouths shut and only tell stories of the Igbo Robin Hood Pirate Cowboy Man who took what was owed to him and shared the wealth."[53] He enacts a further political critique of the oil economy when he tells his daughter, "Those Americans were lucky that I chose to spill their pink blood instead of their red blood. Crush necks of steel instead of flesh. Those goddamn Americans. Like vampires, even in the Nigerian sun."[54] The pink oil is indeed the blood of an American petroleum century, the pipeline a neck of steel holding together the global petroleum infrastructure responsible for hydrocarbon modernity. In these two stories, Okorafor's use of the terms *zombies* and *vampires* recalls Marx's characterization of capital as monstrous. In his analysis of Marx's use of the vampire metaphor and political economy, Marc Neocleous writes, "The vampire, as a 'monster' is of course connected to the root of that term: from *monstrare*, meaning 'to show forth,' *monstra*, meaning 'to warn or show,' *monstrum*, meaning 'that which reveals,' or 'that which warns,' and *monere*, meaning 'to warn.' The vampire as monster both *demonstrates* the capabilities of capital and acts as a *warning* about it."[55] The modality between showing and warning is an apt description of Okorafor's narratorial voice in these two stories. The stories directly address American readers, warning them to see these monsters and be afraid: not only is the monster coming (ecologically speaking and via the logics of predatory capital), but by further tracing petroleum consumption to its extraction sites, the monster is you. These two tropes of the speculative—the monstrous and the apocalyptic—enable the stories' ecocritical interventions.

The stories thus move beyond the dystopic conditions and toward the depiction of US-dominated capitalism as vampiric and monstrous, as that which is revealed by the apocalyptic setting. As David Jefferess notes in his review of *Curse of the Black Gold*, "Oil itself is not the curse or the crisis; structural violence, human rights abuses, and ecological degradation are a consequence of the demand for cheap oil by a certain group of people . . . at the expense of another group of people."[56] Okorafor's writing features the local knowledge that critiques global capitalism, just as Bob Marley did: "Babylon system is the vampire, falling empire . . . suckin' the blood of the sufferers."[57] Marley's musical message of rebellion resonates with the critical lyric message of Fela Kuti, whose song "Zombie" opens Okorafor's story "Spider the Artist," signaling also the analytic power of Black popular music in this work. The epigraph preceding the story cites Fela's lyrics: "Zombie no go stop, unless you tell am to stop / Zombie no go turn, unless you tell am to turn / Zombie! / Zombie no go think, unless you tell am to think."[58] This musical critique travels across a diasporic milieu—Fela Kuti found inspiration in freedom struggles throughout the diaspora, and travel to the United States helped further convince him of the potential of political music; he saw "music as a weapon," explaining that "musicians should be using music to find out what is wrong in the establishment."[59] Okorafor continues this Black Atlantic network of artistic resistance and critique.

The liberatory potential of Black popular music is one example of Okorafor's refusal to depict African settings as dystopic sites of abjection or lack of futurity. In these stories, being Pipeline People involves not only a critical knowledge of the rapacious logics of multinational capital but also a critical engagement with speculative futures and a shaping of those futures toward African ends. In "The Popular Mechanic," Anya's father uses his cybernetic arm to exact a redistribution of oil wealth, however brief. In "Spider the Artist," Eme creates a relationship with the robot that polices the pipeline, in effect repurposing the technology for her own sociopolitical ends, toward her own futurity. Eme escapes her husband's violence by retreating to the backyard, dangerously close to the pipeline, and playing her guitar. Through her musical abilities and her story, she effects a resistance—when she plays her guitar near the pipeline, a certain Zombie repeatedly comes to listen. This Zombie has eight legs and "looked like a giant shiny metal spider. It moved like one, too. All smooth-shifting joints and legs."[60] The spider-zombie is attracted by Eme's guitar playing and returns repeatedly to her backyard, seeming to listen with pleasure to her music. At a moment when Eme expresses despair to the Zombie, playing Marley's "No Woman No Cry" and saying, "My life is shit," the Zombie begins to create its own shiny wire guitar and plays music for her, "a medley of my favorite songs, from Bob Marley to Sunny Ade to Carlos Santana."[61] The Zombie's music at least temporarily restores the natural world: "For a moment all I could hear was the sounds of crickets and frogs singing, the breeze blowing in the palm and mangrove tree tops. I could smell the sizzling oil of someone frying plantain or yam nearby."[62] This musical communion allows the nourishing and sustaining scent of oil for cooking food to temporarily replace the toxic and acrid smell of the oil that otherwise dominates these texts.

The musical relationship between the Zombie and Eme also restores balance to the social world, as Eme's husband stops beating her: "My husband could not beat me when there was beautiful music sending his senses to lush, sweet places. I began to hope. To hope for a baby. Hope that I would one day leave my house and wifely duties for a job as music teacher at the elementary school. Hope that my village would one day reap from the oil being reaped from it. And I dreamt about being embraced by deep blue liquid metal, webs of wire and music."[63] The musical interchange between the AI Zombie and the human narrator presents a way out of gendered violence, a way into creative expression and futurity, in addition to the hope for political and economic justice for the region. Significantly, Eme names her Zombie companion Udide Okwanka, meaning "spider the artist," since "according to legend, Udide

Okwanka is the Supreme Artist. And she lives underground where she takes fragments of things and changes them into something else."[64] This transformational underground form of art and creativity is depicted in the story as a source of change, futurity, and hope. In these ways it is akin to the radical imagination that Robin D. G. Kelley sees as necessary for social change, an imagination accessed through both collective political struggle and "paying attention to the ecstatic."[65]

But Zombies will be Zombies, and soon after the musical Zombie and Eme begin communing, other Zombies "go rogue" and violently turn on the oil workers and the government soldiers working near the pipelines. Eme's musical dream of a "deep blue liquid metal" is juxtaposed to the pink diesel fuel spilling from a broken pipeline. Toxic fumes assault her nose and mouth: "My eyes watered and my nose started running. I held my shirt over my nose and mouth. This barely helped."[66] Although they are programmed to protect the pipeline, the rogue Zombies respond to the bunkering at the broken pipeline by creating an explosion that incinerates the elementary school, many children, and Eme's husband, Andrew, in the process. Eme's Zombie friend, Udide, continues its role of protector, however, shielding her with a force field that renders Eme the sole survivor of the blast. Here Okorafor most clearly subverts a common popular depiction of zombies. This Zombie is capable of sharing, empathy, and creativity. Okorafor's citation of Fela Kuti's song, then, serves as a reminder that the soldiers are not "zombies" either but are people who can choose to turn away from unthinking violent service to a corrupt regime. In Okorafor's depiction, Zombies are not the ever-ravenous undead of popular tropes. Eme credits Udide with allowing her to finally become pregnant, and the two of them serenade her unborn baby daily. Eme wonders what kind of world will exist for her child and again addresses the reader with a warning and a demand: "Pray that Udide and I can convince man and droid to call a truce, otherwise the delta will keep rolling in blood, metal, and flames. You know what else? You should pray that these Zombies don't build themselves some fins and travel across the ocean."[67] The narrator's warning and direct address across the Atlantic calls attention to the connections between the petro-violence in the Niger Delta and the global circuits of petroleum production and consumption. So too does Okorafor's depiction of the machinic music shared between the Zombie Udide Okwanka and Eme.

Kodwo Eshun's groundbreaking music writing in *More Brilliant Than the Sun* describes what he calls sonic fiction, "the hitherto ignored intersections of sound and science fiction."[68] In his introduction, Eshun explains that "the music of Alice Coltrane and Sun Ra, of Underground Resistance and George Russell, of Tricky and Martina, comes from the Outer Side. It alienates itself from the human; it arrives from the future. Alien Music is a synthetic recombinator, an applied art technology for *amplifying the rates of becoming alien*."[69] In "Spider the Artist," music bears this transformative potential. Eme's robotic companion plays an alien, machinic music on a wire instrument it creates. The Zombie's sound is also, to use Eshun's term, "sampladelic," as it plays a recognizable medley of Sunny Ade, Bob Marley, and Carlos Santana for Eme. But after those familiar sounds, "its music deepened to something so complex and beautiful that [Eme] was reduced to tears of joy, awe, [and] ecstasy."[70] The android/Zombie ultimately performs an *alien music* that generates a profound emotional communion with the narrator, who describes it as "something so intricate, enveloping, intertwining . . . *Chei*! I felt as if I was communing with God. *Ah-ah*, this machine and me. You can't imagine."[71] What form of transcorporeality might be embodied in this musical exchange? Is Eme's dream of "being embraced by deep blue liquid metal, webs of wire and music" an expression of another form of liberatory potential within music? Writing of electronic music in 1998, Eshun says, "You are willingly mutated by intimate machines, abducted by audio into the populations of your bodies. Sound machines throw you onto the shores of the skin you're in. The hypersensual cyborg experiences herself as a galaxy of audiotactile sensations."[72] The narrator of "Spider the Artist" has indeed mutated into the Zombie's musical exchange, learning its songs and playing them for her husband. Through playing these songs, Eme gains the agency to literally engineer him toward more peaceful behavior: "My husband hadn't laid a heavy hand on me in weeks . . . ; he was hearing songs that I knew gave him a most glorious feeling. As if each chord, each sound were examined by scientists and handpicked to provoke the strongest feeling of happiness."[73] Thus these two stories present female protagonists who confront the violent environments of their petroleum-ravaged communities and then craft modes of resistance, pleasure, and possibility. While Eme plays beautiful music on her guitar, hoping to be a music teacher, Anya taps trees for palm wine and hopes to be a surgeon. Okorafor's diasporic authorship tells stories of daily local pleasures amid and despite oil's treacherous infrastructure, while her characters display networked knowledges of global oil's larger story.[74]

One of the subversive acts of Okorafor's fiction is the expansion of the "we" who are pipeline people. Connecting the Niger Delta to the sites where oil is consumed undoes the dissociation between energy extraction and energy consumption that might afford some people the luxury of not seeing the pipeline. Despite the difference between "oil people" and "pipeline people"—those reaping profit versus those whose bodies are literally conduits for the conveyance of that profit—ultimately, we are all pipeline people, whether we can see the pipeline or not. Scholarship on oil extraction, read as and alongside ecocriticism, warns that we are all subject to this proximity to oil, inhabiting the dangerous climate of "neoliberal risk, a lethal product of cutthroat corporate cost-cutting, the collapse of government oversight and regulatory authority and the deepening financialization and securitization of the oil market."[75] This climate of risk connects the Niger Delta to the US Gulf, as Okorafor, a US-born writer, draws on the located epistemologies of her parents' homeland and its connection to US petrocapitalism. With an ecocritical diasporic vision, her speculative storytelling posits alternative futures of creative agency.

Notes

1. Before petroleum, palm oil was also a globally traded commodity: "The Niger Delta was integrated into transatlantic commerce from the 15th century onwards," with Europeans trading salt, spices, textiles, slaves, and palm oil. It wasn't until the 1950s that petroleum became the dominant resource extracted from the Delta. Cyril Obi and Siri Aas Rustad, "Introduction: Petro-violence in the Niger Delta—The Complex Politics of an Insurgency," in *Africa Now: Oil and Insurgency in the Niger Delta; Managing the Complex Politics of Petro-violence*, eds. Cyril Obi and Siri Aas Rustad (London: Zed Books, 2011), 211; Michael J. Watts, "Oil Frontiers: The Niger Delta and the Gulf of Mexico," in *Oil Culture*, ed. Ross Barrett and Daniel Worden (Minneapolis: University of Minnesota Press, 2014), 198.

2. Hope Wabuke, "Nnedi Okorafor Is Putting Africans at the Center of Science Fiction and Fantasy," *The Root*, December 29, 2015, https://www.theroot.com/nnedi-okorafor-is-putting-africans-at-the-center-of-sci-1790862186.

3. See, for example, André M. Carrington, *Speculative Blackness: The Future of Race in Science Fiction* (Minneapolis: University of Minnesota Press, 2016); Alondra Nelson, "Introduction: Future Texts," *Social Text* 71, no. 20 (Summer 2002): 1–15; and Lisa Yaszek, "Afrofuturism, Science Fiction, and the History of the Future," *Socialism and Democracy* 20, no. 3 (November 2006): 41–60.

4. Yaszek, "Afrofuturism, Science Fiction, and the History of the Future," 59.

5. Nnedi Okorafor, "Interview with Nigerian Newspaper, *The Daily Trust*," *Nnedi's Wahala Zone Blog*, September 24, 2016, nnedi.blogspot.com/2016/09/interview-with-nigerian-newspaper-daily_3.html.

6. Nnedi Okorafor, “Spider the Artist,” in *Kabu Kabu: Stories* (Germantown, MD: Prime Books, 2013), 102.

7. Reynaldo Anderson, “Critical Afrofuturism: A Case Study in Visual Rhetoric, Sequential Art, and Postapocalyptic Black Identity,” in *The Blacker the Ink: Constructions of Black Identity in Comics and Sequential Art*, ed. Frances Gateward and John Jennings (New Brunswick: Rutgers University Press, 2015), 183. Sandra Jackson and Julie E. Moody-Freeman add that “Afro-Futurism considers issues of time, technology, culture, and race, focusing on Black speculations about the future, foregrounding Black agency and creativity.” “Introduction: The Black Imagination and the Genres,” in *The Black Imagination: Science Fiction, Futurism and the Speculative*, ed. Sandra Jackson and Julie E. Moody-Freeman (New York: Peter Lang, 2011), 3.

8. Kodwo Eshun, “Further Considerations on Afrofuturism,” *CR: The New Centennial Review* 3, no. 2 (Summer 2003): 291–92.

9. Ibid., 292.

10. Ross Barrett and Daniel Worden, “Oil Culture: Guest Editors’ Introduction,” *Journal of American Studies* 46, no. 2 (2012): 272.

11. Studying the BP Deepwater Horizon blowout in the Gulf of Mexico alongside the outrageous rates of spillage in the Niger Delta, for example, Watts notes that “the calamitous events in these two oil gulfs” share “common points of reference in the history of the Black Atlantic.” “Oil Frontiers,” 190.

12. Rob Nixon, “Environmentalism and Postcolonialism,” in *Ecocriticism: The Essential Reader*, ed. Ken Hiltner (London: Routledge, 2015), 200, 203.

13. Rob Nixon, *Slow Violence and the Environmentalism of the Poor* (Cambridge, MA: Harvard University Press, 2011), 273–74 (emphasis mine).

14. Graeme Macdonald, “Oil and World Literature,” *American Book Review* 33, no. 3 (March/April 2012): 31.

15. VèVè A. Clark, “Developing Diaspora Literacy and *Marasa* Consciousness,” in *Comparative American Identities: Race, Sex, and Nationality in the Modern Text*, ed. Hortense J. Spillers (New York: Routledge, 1991): 40–61.

16. Obi and Rustad, “Introduction,” 3.

17. Justina Adalikwu-Obisike and Ebere E. Obisike, “Communities at Risk: An Aftermath of Global Capitalism,” *European Scientific Journal* 10, no. 25 (September 2014): 15–24; Michael J. Watts, “Petro-violence: Some Thoughts on Community, Extraction, and Political Ecology” (Berkeley Workshop on Environmental Politics Working Papers, WP 99-1, Institute of International Studies, University of California, Berkeley, 1999), https://escholarship.org/uc/item/7zh116zd; Obi and Rustad, “Introduction,” 1–14.

18. Obi and Rustad, “Introduction,” 4.

19. Watts, “Petro-violence,” 15. See also Ken Saro-Wiwa, *A Month and a Day: A Detention Diary* (New York: Penguin, 1995); and Nixon, *Slow Violence*, chap. 3.

20. Stefan Skrimshire, “Introduction: How Should We Think About the Future?,” in *Future Ethics: Climate Change and Apocalyptic Imagination*, ed. Stefan Skrimshire (London: Continuum, 2011), 4 (emphasis in original).

21. Alistair McIntosh, foreword to Skrimshire, *Future Ethics*, ix.

22. Walida Imarisha, introduction to *Octavia’s Brood: Science Fiction Stories from Social Justice Movements*, ed. Adrienne Maree Brown and Walidah Imarisha (Oakland, CA: AK Press, 2015), 4.

23. Robin D. G. Kelley, *Freedom Dreams: The Black Radical Imagination* (Boston: Beacon Press, 2002), 191.

24. John Joseph Adams, “Author Spotlight: Nnedi Okorafor,” *Lightspeed Magazine* 10 (March 2011), www.lightspeedmagazine.com.

25. Imre Szeman, “Introduction: Pipeline Politics,” *South Atlantic Quarterly* 116, no. 2 (April 2017): 402; Watts, “Oil Frontiers,” 193.

26. Szeman, “Introduction,” 403.

27. Graeme Macdonald, "Containing Oil: The Pipeline in Petroculture," in *Petrocultures: Oil, Politics, Culture*, ed. Sheena Wilson, Adam Carlson, and Imre Szeman (Montreal: McGill-Queen's University Press, 2017), 39.

28. Nnedi Okorafor, "The Popular Mechanic," in *Kabu Kabu: Stories* (Germantown, MD: Prime Books, 2013), 167.

29. Okorafor, "Spider the Artist," 101.

30. See Obi and Rustad, "Introduction"; Stephanie LeMenager, *Living Oil: Petroleum Culture in the American Century* (New York: Oxford University Press, 2014), 126–31; and Nixon, *Slow Violence*, chap. 3.

31. Okorafor, "Spider the Artist," 101.

32. Ibid., 102.

33. Ibid., 104. *Oyibo* means "white person" in Nigerian pidgin.

34. Ibid. (emphasis mine).

35. Ibid., 105.

36. Ibid., 105–6 (emphasis mine).

37. Obi and Rustad, "Introduction," 10.

38. LeMenager's chapter, "Petromelancholia," in *Living Oil* analyzes two realist novels about petroleum, one set in the Niger Delta and one set in Houston, Texas. See esp. 125–37, in which she analyzes two authors, writing "across two oceans into the heart of what has been called by the oil industry, the New Golden Triangle (West Africa-the Gulf of Mexico-Brazil)," 136.

39. Okorafor, "Spider the Artist," 106 (emphasis in original).

40. Adalikwu-Obisike and Obisike, "Communities at Risk," 20.

41. Ibid., 17.

42. Obi and Rustad, "Introduction," 4. Clearly, the US state also acts in partnership with oil companies. As early as 1973, a National Science Foundation report found "widespread collusion between industry and government and very little government oversight. The U.S. Geological Survey freely granted waivers from complying with the limited regulations and inspection demands." Watts, "Oil Frontiers," 204.

43. Obi and Rustad, "Introduction," 10.

44. Okorafor, "Spider the Artist," 101.

45. Okorafor, "Popular Mechanic," 167. It is important to note that despite repression, pipeline sabotage has also been effective in closing some aspects of oil MNC operations in the Niger Delta. Watts, "Oil Frontiers," 197.

46. Okorafor, "Popular Mechanic," 166–67.

47. For example, a lawsuit was brought against Pfizer for its 1996 drug testing in Nigeria. George J. Annas, "Globalized Clinical Trials and Informed Consent," *New England Journal of Medicine* 360, no. 20 (May 14, 2009): 2050–53.

48. Okorafor, "Popular Mechanic," 167.

49. Ibid., 170.

50. Ibid., 171.

51. Ibid., 172.

52. Ibid., 174.

53. Ibid.

54. Ibid., 174–75.

55. Mark Neocleous, "The Political Economy of the Dead: Marx's Vampires," *History of Political Thought* 24, no. 4 (Winter 2003): 684 (emphasis in original).

56. David Jeferess, "Oil's Long Shadow," *American Book Review* 33, no. 3 (March/April 2012): 5.

57. Bob Marley, "Babylon System," *Survival*, Tuff Gong Recording Studio, 1979.

58. Okorafor, "Spider the Artist," 101.

59. George Lipsitz, *Dangerous Crossroads: Popular Music, Postmodernism, and the Poetics of Place* (London: Verso, 1994), 140.

60. Okorafor, "Spider the Artist," 105.

61. Ibid., 109.

62. Ibid., 108–9.

63. Ibid., 110.

64. Ibid., 111. Okorafor states, "In West African culture, spiders tend to represent creativity and storytelling." John Scalzi, "The Big Idea: Nnedi Okorafor," *Whatever* (blog), June 17, 2010, https://whatever.scalzi.com/2010/06/17/the-big-idea-nnedi-okorafor/. Okorafor's 2014 novel *Lagoon* also refers to Udide Okwanka. *Lagoon* (New York: Saga Press, 2016), chaps. 44 and 57.

65. Kelley, *Freedom Dreams*, 10.

66. Okorafor, "Spider the Artist," 113.

67. Ibid., 115.

68. Kodwo Eshun, *More Brilliant Than the Sun: Adventures in Sonic Fiction* (London: Quartet Books, 1998), 10.

69. Ibid., 8 (emphasis mine).

70. Okorafor, "Spider the Artist," 109.

71. Ibid., 111.

72. Eshun, *More Brilliant*, 10.

73. Okorafor, "Spider the Artist," 110.

74. *Lagoon* features a female marine biologist who meets shape-shifting extraterrestrial creatures sabotaging offshore oil rigs. I develop a reading of that novel's ecocritical lens and Black Atlantic musical focus in Wendy W. Walters, "Seascapes of Afrofuturism" (paper, American Comparative Literature Association, University of Utrecht, July 2017).

75. Watts, "Oil Frontiers," 196.

6.

Petro-drama in the Niger Delta
Ben Binebai's *My Life in the Burning Creeks* and Oil's "Refuse of History"

Henry Obi Ajumeze

Crude oil was a mix of hydrocarbons and impurities ill-suited for most purposes: it burned unsteadily, released large amounts of soot, and often smelled foul.
—Christopher F. Jones, *Routes of Power: Energy and Modern America*

Joshua Esty has coined the term "extremental postcolonialism" to trace the significance of scatology in the emergence of first-generation African writing and to extensively theorize the artistic imagination of dystopia that heralded the postcolonies.[1] To be sure, Esty's figuration of scatology productively gestures toward the ways that postcolonial discourses are extended to implicate different forms of power formations—including the power of odor. Drawing on Dain Borges's phrasing of "belly politics,"[2] which anticipates Achille Mbembe's "aesthetics of vulgarity"[3] in which the postcolony is described as a site of obscene magnificence, Esty notes that "the remarkable currency and symbolic versatility of excrement in the postcolony" is an affirmation of failed or flawed postcolonial nationalism.[4] To invent shit as a conventional portrait of damaged nationhood, therefore, is the product of a cultural response to politics that dates back to postindependent African writing. Indeed, Mbembe finds the operations of vulgarity in the postcolony to be "intrinsic to all systems of domination and to the means by which those systems are confirmed or deconstructed."[5] Not surprisingly then, succeeding generations of African writers have continued to invest in the language of shit to deconstruct the rapacity of postcolonial plunder. In particular, writers imagining the pollution of the landscapes of the Niger Delta through extractive activities have drawn readers into varied contexts in which the environment is rendered in the form of material excrescence. In their exploration of the ecological devastation of oil modernity in the region, these writers perform the task that Rachele Dini describes as telling "stories in which human beings are classed as worth less than trash."[6]

In this chapter, I argue that oil tropes in literature are found with a particular kind of dystopia in Nigeria and thus foster a new epoch of politics that announces state failure most evidently through the manifestation of ecological desecration. In other words, I suggest that rethinking scatology in the expanded time and era of petro-politics works to mirror the sense of failure and violence that the circuits of extraction make possible.

It is Amitav Ghosh, in "Petrofiction"—his well-known review of Abdelrahman Munif's novel *Cities of Salt*—who first indicates how the material abstraction of oil culminates in a kind of putrefaction. In setting out to examine why the bourgeoning economic and political value of oil in the twentieth century has failed to nurture great fiction—in comparison with the spice trade, which demonstrated equivalent historical significance in the sixteenth century—Ghosh identifies the central reason as the manner in which oil resonates in the system of global capitalism. On the question of oil's implication in culture, he asserts that "the smell of oil gets a lot worse by the time it seeps into those rooms where serious fiction is written and read. . . . And to make things still worse, it begins to smell of pollution and environmental hazards. It reeks, it stinks."[7] Although Ghosh theorizes that the experience of oil reproduces a sense of placelessness that accords with physical absence, the scatological imagination that he inaugurates is useful in addressing the character of oil in material terms. In other words, an important aspect of measuring oil's relation with culture is to acknowledge its physicality rather than its invisibility—an acknowledgment that speaks to the reality of oil's presence at the site and environment of extraction. The Nigerian scholar Philip Onoriode Aghoghovwia coined the phrase "poetics of cartography"[8] to register the effect of oil encounters on the local ecology of the Niger Delta, arguing that the geographic particularities of resource extraction and distribution are significant in coming to terms with the ontology of oil. In his view, the literary response to oil culture in the Global South considers "the material texture of the oil's presence in the local communities."[9] Amy Riddle has argued that the detailed physical description of oil is central to its contemporary representation—in contrast with earlier times represented by *Cities of Salt*—of what underscores the growing experience of globalism.[10] To put it another way, oil's registration in human culture is framed by different historical circumstances that manifest themselves in the Global South in the framework of what Aghoghovwia describes as "physical, spatial intrusion."[11] Indeed, the physical form of crude oil continuously forces its visibility in the landscape of extraction to reveal tendencies that define different shades of cultural imagination. As Riddle acknowledges, "The supra-objective qualities of oil as both fuel and plastic, earth and air, subject and system, distinguish it from earlier commodities in literature, like coffee, spices, or sugar."[12] In an effort to depart from the notion of oil's invisibility by emphasizing its physical character in the landscape of extraction, I argue here that this distinctiveness finds expression in how tropes of waste and decay of crude oil matter are imagined and performed. How might one understand the artistic imagination of decay implicit in oil's physical properties as a kind of violence and resistance against the oppressive protocols of multinational oil corporations?

Reimagining a geographical violence through the lived cultures of oil in Nigeria's Niger Delta, Ben Binebai's stage play *My Life in the Burning Creeks* (2010) presents the frictions within petromodernity that echo "oil's dirtiness and fecal qualities."[13] Drawing attention to instances of state repression and securitization in oil-producing communities, the play highlights the creeks of the Niger Delta as landscapes of intense dramatic action, particularly as the fulcrum of the region's history of petro-violence. It registers the land in which oil was discovered as a landscape of material waste and rot, one that is constantly violated with consequences of underdevelopment and destruction of living environments. To put this into context, when the Nigerian Conservation Foundation convened international experts to assess the damage caused by petro-extraction in 2006, the report emphasized ecological desecration of the Niger Delta, noting that the region is "one of the world's most severely petroleum-impacted ecosystems and one of the top five most polluted places."[14] This notion of pollution in the Niger Delta is imagined in ways that describe the ruin of a "land of beauty" through oil spillages by multinational petroleum corporations. The play's eponymous narrator, Pereware, echoes thus in the play's prologue:

> Our generation has risen against the
> Continuous oil exploration without
> Corresponding development of our land.
> The struggle against corporate irresponsibility,

State corruption, balkanization, poverty,
Neo-colonization and bio-regional damage
Has transformed our land of beauty
Into a roaring and burning Zone.[15]

The play opens with an air strike on Gbaramatu Kingdom, one of several oil-producing communities in the Niger Delta. The region is reproduced as a "theatre of operation" in which the Nigerian security system wages wars on the local inhabitants. Binebai grounds the plot in a real-life event that occurred in 2009: air bombardment of the palace of the Gbaramatu Kingdom by the Joint Task Force (JTF). The bombardment was ordered by the Nigerian federal government in connection with the hunt for some members of the Movement for the Emancipation of the Niger Delta (MEND)—one of the insurgent groups that operated in the region during the period. The journalist Simon Ebegbulem, reporting the event as a manifestation of the politics of oil and terror, compared the offensive with other familiar instances of violence and precarity in petro-states. He wrote that "the Gbaramatu invasion by the JTF was similar to the operation Desert Storm of [1990–91], when American forces launched an attack against Iraq for invading Kuwait."[16] The bombing is evocative of a pattern of reprisal attacks on some Niger Delta villages, such as in the case of Odi town, in which hundreds of people were massacred in 1999. In reporting the Odi Massacre, the British writer Michael Peel observed that the Nigerian president at the time, Olusegun Obasanjo, "showed no signs of outrage at the action and others like it"[17] but rather boasted that the vengeful massacre by security forces is a typical example of how cause and effect function in human societies. Such is the violent strategy that frames the politics of oil in the Delta, a pattern of normalized securitization that the play orchestrates. As Frederick Buell writes, "Oil has become an obsessive point of reference in and clear determinant over the daily lives of many, either victimizing them directly and cruelly . . . or making them increasingly feel that their developed-world normalities are a shaky house of cards."[18] The play's narrator relives these cruel experiences not only to illustrate the exclusion of the oil-bearing communities from social and economic development but also to highlight ways in which they are exposed to what Rob Nixon describes as "modernity's false dawn."[19] It captures the narrator's life several years after university graduation—although he has a degree in petroleum engineering, he is unable to secure gainful employment in any of the multinational oil companies based in his native oil-producing community. He asks, "Is it not sad that after a decade / Of graduation from one of the most prestigious Nigerian universities / One remains unemployed?"[20] The narrator's life contrasts sharply with that of his friend Abubakar from the northern region of Nigeria, who got a degree in Islamic and Arabic Studies yet easily secured a place at the public and corporate affairs unit of Escravos's Chevron, one of the top oil companies in the region. The story resonates with what scholar James Tsaaior describes as a "decimal of a tissue of paradoxes" in which indigenes of the oil-rich Delta live "on the margins or fringes of Nigeria's national life courtesy of perennial institutional and state neglect."[21] This "idiom of north/south dichotomy"[22] has become an important reference point in the discourse of oil culture in the Niger Delta literature, as it articulates a broad set of power relations useful in thinking about the geopolitical circuitry of production and consumption of oil in Nigeria. Drawing on the production sites of the American mid-Atlantic region, Christopher F. Jones gives further insight into this geopolitical perspective of oil:

> The production of oil came at great environmental cost. But users who lived hundreds and even thousands of miles away did not need to worry about oil flowing into their streams or ruining their soil. They had little personal connection to the massive deforestation of large regions in pursuit of liquid gold. Even if they experienced some soot and odor from the burning of kerosene in their homes, it was likely cleaner and more pleasant than candles or camphene. For the most part, the users of oil gained the benefit of cheap energy without assuming responsibility for its environmental damage. One of the reasons that fossil fuel energy production has been so environmentally destructive is that those who benefit from energy sources rarely have to live with the environmental damages associated with its production.[23]

In the Niger Delta scenario, the geopolitics of oil is further exacerbated by the minority status of the oil-producing region within the larger Nigerian state. In this light, the struggle for resource control is undertaken not only to instrumentalize fiscal federalism, which is skewed by the systems of power that control the rentier economy of oil, but also to serve as a strategy to redefine the protocols of resource entitlement. Not surprisingly, then, the play instantiates a heightened sense of artistic aggression that recognizes and echoes the campaign for self-determination by the Niger Delta youth militias: "The militants face the oppressors / Face to face, hand to hand and / Dance their smothering and vicious guns / To stillness."[24] When the air raid occurred—in the form of a reprisal attack—Pereware is still hunting for a job and is trapped in the mass of displaced and dispossessed people in the creeks. His lament resonates with the abjection and bare life to which Deltans are reduced:

> PEREWARE: But Abubakar, how do I make it
> When I don't have a Godfather?
> I need the job, would you take me
> To a godfather who can offer me the job?
> PEREWARE as ABUBAKAR: The bitter truth is that you
> Need to become a Muslim
> And truly bear a Muslim name before
> The assistance can come.
> PEREWARE: What? I did not bargain in life
> To change my religion
> And biological identity because of a job[25]

Given the play's emphasis on the ruination of the Delta landscapes through the activities of the Nigerian state and multinational petroleum companies, Binebai is revealed as a dramatist whose artistic logic is defined by the material consequences of petro-violence on the natural environment. He stages conflicts that acknowledge the centrality of the natural environment in the precarious conditions of devastation and pollution. These conflicts invoke the landscape in order to expose the realities of petro-violent predation by the Big Oil corporations and to reflect on the contrast with the premodern period, during which the landscape was the source of the Delta's economic and agrarian livelihood. The play's revelations about the destruction of the region's landscapes and about Pereware's quotidian travails demonstrate a tragedy common to man and land in the context of decay. The idealization of the pre-oil Delta as evident here in the narrator's appellation offers insight into the degree of tragic transformation that the oil encounter wrought in the Delta region.

> The king without a crown
> The emperor without clothes,
> The peacock without feathers and like a
> Bloated cash cow in the trade

Garden of some capitalists
This land of mine hosts fishing grounds
And farm lands that diminish everyday as
The liquid gold slicks flicker the creeks
And rivers burst into
Dazzling and deafening flames[26]

The play points to the ways in which the impact of oil extraction subjects the creeks of the Niger Delta to a bare life of destruction, offering a story about Indigenous people in a fractured seascape and the complex overdetermination of resistance that emerges from the relation to state terror. Turning the region into a wasteland of detritus and encompassing all sorts of spatial crises, oil extraction inscribes material destruction in the waves of petromodern activities in the region. To be sure, the creek is the agricultural heartland of Nigeria, imagined in popular culture as the food basket of the nation, such that subjection to resource extraction instantiates crude violence and dispossession. In this context, the creeks reflect the complex ways in which a sustainable livelihood, the natural environment, and toxic politics reinforce one another while projecting new ways of aestheticizing the political ecology of oil. The evident dramatization of the culture of waste and pollution in what underwrites a poetics of scatology dominates *My Life in the Burning Creeks*, culminating in a spatio-material framework in the play's enactment of petroviolence. In other words, Binebai makes the creeks the melting point of different kinds of pollution that are traceable to the culture of oil:

This oil that has polluted
The burning creeks
Has also polluted politics and democracy
It has polluted national unity
It has polluted the lives of youths
It has polluted our rulers, elders, girls
And has turned into a colossal catalyst
For crooked politicians and mercantilists.[27]

In *A Swamp Full of Dollars*, Michael Peel gives an account of his visit to the oil communities in the Niger Delta and notes that infrastructural decay and abandonment is an important lens through which to view Nigeria's lopsided federal system. His description of Oloibiri, the historical town where crude oil was first discovered in the region in 1956, particularly elaborates on the overarching culture of deprivation in Nigeria's petrocultural modernity. Ike Okonta and Oronto Douglas also observe Shell's culture of ecological pollution and devastation of the Delta region's biodiversity through practices ranging from the unregulated use of dynamite to prospect for oil to the "obnoxious practice of gas flaring and the oil leakages of Shell's old, rusty and corroded [pipelines]."[28] Environmental movements have blamed pipeline fire disasters on the poor condition of oil facilities in the region and on technologies of oil that are in various stages of disrepair, rusted and corroded. In the words of the environmentalist Nnimmo Bassey, Shell is renowned for its global indifference to the territories of host countries and has a history of turning "pristine wetland into wastelands."[29] Peel's reflection on Oloibiri offers important insight into the condition of local ecologies that are rendered bare after extraction:

> As I approached Oloibiri one typical Delta morning already thick with sticky heat, I passed an anonymous-looking clearing hacked from the jungle. A rusty barbed wire fence surrounded an equally dilapidated oil-wellhead, known in the industry as a Christmas tree because of its branching network of pipes and valves. The accompanying signboard, rendered barely legible by corrosion, was as understated a historical landmark as you could wish to find . . . anywhere: "Oloibiri: well number one," it read. "Drilled June 1956. Depth, 12,008 feet."[30]

Binebai echoes this narrative of decay and abandonment to suggest ways in which the material presence of oil brings about rust and corrosion that erase the historical landmarks of the region. In other words, oil's physicality is evident in the destructive sense that reveals lack of careful and measured production that is eventually submitted to disuse. This fact resonates with Christopher F. Jones's reflection on the liquid nature of oil, in which he illustrates the difficulty of controlling the flows as it "seeped out of wells, ran into streams, and leaked out of wells."[31] He notes that "the materiality of oil, combined with human disregard for spills, played a significant role in the sacrifice of the environment of the Oil Regions."[32] Considering the rust that attends to signposts of oil in Oloibiri ("the sorry state of the signpost: the fast fading inscriptions, the rusty board"), Aghoghovwia suggests the consideration of rust and decay as a metaphor for invisibility: "Isn't this symptomatic of the invisibility of Oloibiri—and perhaps the entire Niger Delta—in the context of social development?"[33] If, as Bruno Latour has argued, "visibility is the consequence of lots of opaque and 'invisible' work,"[34] then the condition of invisibility to which the Delta is subjected suggests greater opacity embedded in political conspiracies in the region. In presenting the trope of decay, *My Life in the Burning Creeks* appears to use the narrative strategy that Walter Benjamin styles as "literary montage" to ask questions of the broken territories and silent topographies of the Delta. In *Arcades Project* (*Passagen-Werk*), Benjamin deploys the theory of montage as a methodology of historical writing. He explains the technique thus: "I have nothing to say. Only to show. I will steal no valuables, nor appropriate any clever turns of phrase. But the rags, the refuse: not in order to take stock of them but to use them—which is the only way of doing them justice."[35] To be sure, Benjamin's obligation to merely show the rags and "the refuse" of history as a way of seeking justice parallels Binebai's dramatic representation of the ruination and despoliation of the Delta landscape. Although Binebai indulges in "clever turns" in which he romanticizes the Delta landscape in order to hold the mirror up to the refuse that history has made of the region, the suggestion of the theatrical details through which the play turns its gaze on the ruination underscores a quest for environmental justice for the predation wrought by petro-capitalism. Hence, when Pereware invokes imaginations of environmental history to tell stories of the decay and death of place and geography, it is to provoke what Benjamin ascribes to the montage methodology as "the technique of awakening." Such an assumption of the audience's socioenvironmental awakening finds expression in the play's reinstatement of the geophysical and spatial affect of the oil experience as shown in the narrator's lamentation:

> I went dry like Oloibiri, Burutu, Ganagana
> And Forcados, prosperous ancient communities
> In the burning creeks, used and
> Dumped by the European traders in the 20th century
> I attended primary and secondary
> Schools at Burutu, fondly called
> The Island of No regrets.
> It was home to all Nigerians
> For more than seventy years
> This Island town was the nerve centre
> Of commerce and industry throughout

British West Africa
Burutu the Island town, the port
Town with the largest slipways,
The Island that is designated as the ideal base for
The largest ship building and
Repairing centre in Africa,
Is now a shadow of its old self like
Forcados which served as the first capital of
Colonial Southern Nigeria.
The death of these towns of industry and
Economic hub is a monumental blow to
Our hopes of survival in the burning creeks.[36]

The emphasis on the history of the major towns in the Niger Delta, a place that was geostrategically part of, and later not part of, the Anglo-American relation with its natural resources, makes this play a remarkable geopolitical drama. It unmasks the European interest in a basic question of resource exploitation in which the lands are the material victims. In fact, this geographical mapping of historical Delta towns that are forgotten soon after they are drained of natural resources brings to the play a concrete sense of land politics and underlines its gesture toward an account of the region's environmental history. It seems significant that the play speaks about "the death of these towns," which suggests thinking of tragedy in spatial and territorial terms. *My Life in the Burning Creeks* articulates this notion of landscape death by deploying the trope of decay and rust, a way of confronting the materiality of petroleum in the region with myriad forms of planetary loss and desecration. It might be supposed that what Binebai has done is to draw from the destruction of the oil landscapes and project it into a narrative idiom of scatology. Through Pereware's constant references to excrement in relation to life in the Niger Delta, Binebai appears to echo the former president of Venezuela who famously termed oil as "the devil's excrement." Michael Watts describes this more elaborately as "the popular understanding of petroleum as socially polluting,"[37] which underscores its material effect on the oil-bearing communities "with its power to tarnish and turn everything into shit." Stephanie LeMenager has noted that this character of oil that derives from "the primal association of oil with earth's bodies" often poses representational crises. Binebai's play appears to be a good example of LeMenager's argument about "the permeability, excess, and multiplicity of all bodies deemed performed and given."[38] In the play, shit permeates the operational structures of the oil companies as illustrated in the toilet episode, which becomes useful in exhibiting how the application letters of jobless indigenes of the Delta are trashed. In this instance, while "looking for a soft paper" to clean his anus after defecation during an interview with one of the oil companies, Pereware discovers a photocopy of his master's degree certificate in a heap of paper waste. This heightened sense of abjection of the Delta Indigenes underwrites the neoliberal marginalizing system that is constituted to render life into shit. In other words, the reinforcement of vulgarity in the context of waste exhibits what Slavoj Žižek describes as "systemic violence." For Žižek, systemic violence flows invisibly through the social structures that capitalism has enabled and instituted. In this light, it is "the violence inherent in a system: not only direct physical violence, but also the more subtle forms of coercion that sustain relations of domination and exploitation."[39] Thus, the play makes visible the subtle systems that sustain the different kinds of violence that are unleashed on the oil-bearing communities in the Niger Delta. It provides physical evidence of the interplay of marginalizing power relations by invoking images of shit to parallel "oil's dirtiness and fecal qualities." The account of the narrator's unutilized certificate as detritus is as personal as it is collective in the Delta paradox of oil wealth:

I squatted to ease myself and
When I finished defecating,
I found comfort in looking for a soft paper
To clean my anus. I eventually
Saw one and was going to use
It straight away but I saw
The reverse side of it
I was devastated emotionally
Something strangely familiar . . .
What an irony of fate
The certificate I submitted
For a job in the oil company
Operation in my land was returned to
Me through the toilet.[40]

This kind of representation that locates aesthetics in decay as a way of articulating certain kinds of power relations resonates with Bakhtin's formulation of the official and unofficial binary. In his study of *Rabelais and His World*, Bakhtin reproduces this binary while discussing the imagination of scatological subjectivities in literature—which he terms "grotesque realism"—and defining shit as "the most suitable substance for the degrading of all that is exalted."[41] He thus argues that obscenity functions to degrade the monological disposition of the official cultures. In addition, Bakhtin believes that excremental matter illuminates the laughing aspect of the world such that authority is resisted through the instrumentality of ridicule, producing regenerative and renewing principles that are often glossed over by critics. As he writes, "The images of feces and urine are ambivalent, as are all the images of the material bodily lower stratum; they debase, destroy, regenerate, and renew simultaneously. They are blessing and humiliating at the same time."[42] This view makes possible the discussion of how Binebai invents the fecal matter as a way to express the degradation of the landscape and people of the Niger Delta. *My Life in the Burning Creeks* co-opts the ruin and decay of the ordinary people of the Delta to speak broadly about the material rot of the landscape in which they dwell. Put differently, to speak of decay is to address the industrial predators together with their Nigerian government collaborators.

When Bakhtin invokes the "grotesque body" while emphasizing shit as one of the important products of body matter, it offers an ecologically relational perspective on the discussion of scatology that is useful in thinking about *My Life in the Burning Creeks*. As he puts it, "Dung is a link between body and earth, urine is a link between body and sea."[43] Binebai's interest in the anus as the body's site of waste is symptomatic of the general despoliation and waste of the region as a result of petro-extraction. It offers a way of dramatizing the famous assertion of the Polish journalist Ryszard Kapuscinski that "oil is a filthy, foul-smelling liquid that squirts obligingly up into the air."[44] Such a view is echoed by Christopher F. Jones in the epigraph to this chapter to articulate the biophysical character of oil. This aspect of oil ontology is central in the play more broadly, as it assumes a general metaphor of filth and scatology. Indeed, a heightened sense of filth and malodor finds expression in Pereware's announcement during the funeral of his mother: "I perceived the reeking scent of dipping / my finger into my anus."[45] In keeping with a sense of the ecological lineaments of petro-capitalism, one could say that the play evinces an aesthetic interface that connects the decay of self and the pollution of the environment. That is, the lifescape and landscape of the Delta are implicated in the conditions of decay and pollution that function in a system of relations. Pereware, for instance, declares this about the occupational closure of a Delta fishing community whose subsistence is incapacitated precisely because of the pollution of the rivers: "Becoming a fisherman is futile because oil / Has polluted the waters / What about the flora and fauna / That have been deprived of their fertility?"[46] *My Life in the Burning Creeks* thus explores the trope of scatology in ways that juxtapose human decay with environmental pollution visited on the oil communities in the Delta. It identifies obscenity in the everyday life of Deltans who are economically and spatially displaced in landscapes damaged by extraction. To be sure, Binebai appears to explore environmental pollution by giving expression to the experience of human dispossession. In highlighting the transcendental effect of pollution—its permeability and flow—and giving purchase to its bodily, economic, and political implications, the play enunciates obscenity as a kind of ecological resistance. It articulates the Bakhtinian logic of the grotesque as a mode of resistance, one that acts on the tragic consequences of decay. In fact, the excrement gives profound expression to what Slavoj Žižek calls an "excremental/sacred outcast,"[47] which implies a category of exclusion that Deltans are subjected to outside the limits of economic and political recognition. In this sense, excrement becomes a tool for political agency and subversion as well as a kind of aesthetic paradox that invokes the exclusion and marginalization of the Deltans as outcasts.

Beginning with the title, *My Life in the Burning Creeks* unfolds as an imaginative landscape story that draws the human experience into a narrative common: the creeks and the people who are embedded in them and how these people are inscribed in the cruel and damaging encounter with extractive politics. In this sense, the play is the deleterious account of a commodity that, since its discovery in 1956, permeates and flows through life and that confronts those who live in the landscape of its production with its visible, material presence. What Patricia Yaeger describes as the "inquiry about energy's visibility or invisibility"[48] gravitates toward a conundrum of sorts at the site of resource extraction to reveal how the materiality of oil preeminently erases, or makes invisible, social and political development in the oil region. My reading thinks through the odor of oil to reflect on how it finds representation in decay and filth, affirming a growing culture of scatology that is becoming normalized in most oil-producing communities in the Global South. This perspective, I argue, is an aesthetic abstraction of the biophysical property of the commodity through which materiality is metaphorized. In other words, oil is no more invisible than its ontological presence in the substratum of polluted landscapes in which its damaging smells and decays are tellingly evident. This mode of writing oil offers a kind of subversive art that reflects the mood of resistance against the Nigerian state and the petroleum companies.

Notes

1. Joshua D. Esty, "Extremental Postcolonialism," *Contemporary Literature* 40, no. 1 (1999): 36.
2. Dain Borges, "Machiavellian, Rabelaisian, Bureaucratic?," *Public Culture* 5, no. 1 (1992): 109.
3. Achille Mbembe, "The Banality of Power and the Aesthetics of Vulgarity in the Postcolony," *Public Culture* 4, no. 2 (1992): 1.
4. Esty, "Extremental Postcolonialism," 36.
5. Mbembe, "Banality of Power," 2.
6. Dini Rachele, "'Resurrected from Its Own Sewers': Waste, Landscape, and the Environment in J. G. Ballard's 1960s Climate Fiction," *ISLE: Interdisciplinary Studies in Literature and Environment* 26, no. 3 (2019): 5.
7. Amitav Ghosh, "Petrofiction: The Oil Encounter and the Novel," *New Republic*, March 1992, 30.
8. Philip Onoriode Aghoghovwia, "Poetics of Cartography: Globalism and the 'Oil Enclave' in Ibiwari Ikiriko's *Oily Tears of the Delta*," *Social Dynamics* 43, no. 1 (2017): 32.
9. Ibid., 30.
10. Amy Riddle, "Petrofiction and Political Economy in the Age of Late Fossil Capital," *Mediations* 32, no. 1 (2018): 56.
11. Aghoghovwia, "Poetics of Cartography," 39.
12. Riddle, "Petrofiction and Political Economy," 55.
13. Stephanie LeMenager, *Living Oil: Petroleum Culture in the American Century* (Oxford: Oxford University Press, 2014), 92.
14. Cited in Adati Ayuba Kadafa, "Oil Exploration and Spillage in the Niger Delta of Nigeria," *Civil and Environmental Research* 2, no. 1 (2012): 38–51.
15. Ben Binebai, *My Life in the Burning Creeks* (Ibadan, NG: Temple, 2014), 49.
16. Simon Ebegbulem, "Gbaramatu: A King Without Palace," *Vanguard*, May 28, 2011, 4.
17. Michael Peel, *A Swamp Full of Dollars: Pipelines and Paramilitaries at Nigeria's Oil Frontiers* (Ibadan, NG: Bookcraft, 2010), 10.
18. Frederick Buell, "A Short History of Oil Cultures: Or, the Marriage of Catastrophe and Exuberance," *Journal of American Studies* 46, no. 2 (2012): 274.
19. Rob Nixon, *Slow Violence and the Environmentalism of the Poor* (Cambridge, MA: Harvard University Press, 2011), 42.
20. Binebai, *My Life in the Burning Creeks*, 14.
21. James Tsaaior, "Poetics, Politics and the Paradoxes of Oil in Nigeria's Niger Delta Region," *African Renaissance* 2, no. 6 (2005): 2.
22. Oyeniyi Okunoye, "Alterity, Marginality and the National Question in the Poetry of the Niger Delta," *Cahiers d'études africaines* 48, no. 191 (2008): 420.
23. Christopher F. Jones, *Routes of Power: Energy and Modern America* (Cambridge, MA: Harvard University Press, 2014), 143–44.
24. Binebai, *My Life in the Burning Creeks*, 32.
25. Ibid., 21.
26. Ibid., 19.
27. Ibid., 53.

28. Ike Okonta and Oronto Douglas, *Where Vultures Feast: Shell, Human Rights and Oil in the Niger Delta* (San Francisco: Sierra Book Club, 2001), 196.

29. Tsaaior, "Poetics, Politics and the Paradoxes of Oil," 75.

30. Peel, *Swamp Full of Dollars*, 25.

31. Jones, *Routes of Power*, 121.

32. Ibid.

33. Philip Onoriode Aghoghovwia, "The Drama of Oil Production in the Niger Delta: An Obsession with the Spectacular," *JWTC Blog*, August 15, 2019, https://jhbwtc.blogspot.com/2012/07/drama-of-oil-production-in-niger-delta.html.

34. Bruno Latour and Katti Christian, "Mediating Political 'Things,' and the Forked Tongue of Modern Culture: A Conversation with Bruno Latour," *Art Journal* 65, no. 1 (2006): 98.

35. Walter Benjamin, *The Arcades Project*, trans. Howard Eiland and Kevin McLaughlin (London: The Belknap Press of Harvard University Press, 1999), 574.

36. Binebai, *My Life in the Burning Creeks*, 53.

37. Michael Watts, "Petro-violence: Community, Extraction, and Political Ecology of a Mythic Commodity," in *Violence Environments*, ed. Nancy Lee Peluso and Michael Watts (Ithaca: Cornell University Press, 2008), 212.

38. LeMenager, *Living Oil*, 92.

39. Slavoj Žižek, *Violence: Six Sideways Reflections* (New York: Picador, 2008), 10.

40. Binebai, *My Life in the Burning Creeks*, 23.

41. Mikhail Bakhtin, *The Dialogic Imagination: Four Essays*, ed. Michael Holquist (Austin: University of Texas Press, 1981), 152.

42. Ibid., 151.

43. Ibid., 335.

44. Ryszard Kapuscinski, *Shah of Shahs* (London: Quartet Books, 1982), 152.

45. Binebai, *My Life in the Burning Creeks*, 43.

46. Ibid., 52.

47. Slavoj Žižek, "Slavoj Žižek on His Favourite Plays," interview by Liza Thompson, Five Books, May 2016, http://fivebooks.com/interview/slavoj-zizek-favourite-plays/.

48. Patricia Yaeger, "Editor's Column: Literature in the Ages of Wood, Tallow, Coal, Whale Oil, Gasoline, Atomic Power, and Other Energy Sources," *PMLA* 126 (2011): 137.

7.

Documenting "Cheap Nature" in Amitav Ghosh's *The Glass Palace*

A Petro-aesthetic Critique

Stacey Balkan

"My dear doctor," said Flory, "how can you make out that we are in this country for any purpose except to steal? It's so simple. The official holds the Burman down while the business man goes through his pockets. Do you suppose my firm, for instance, could get its timber contracts if the country weren't in the hands of the British? Or the timber firms, or the oil companies, or the miners and planters and traders? How could the Rice Ring go on skimming the unfortunate peasant if it hadn't the Government behind it?"
—George Orwell, *Burmese Days*

It's too often forgotten that capitalism's energy revolution began not with coal but with wood—and with the privatization that forest enclosure implies.
—Raj Patel and Jason W. Moore, *A History of the World in Seven Cheap Things: A Guide to Capitalism, Nature, and the Future of the Planet*

The "fragile hegemony of extractive capitalism" under which our global petrosphere operates is made possible (in part) through networks of itinerant labor regimes that are strikingly absent from extant discussions around "economic development"—a euphemism that rings particularly hollow in nations such as Myanmar (formerly Burma) where global speculation has decimated local communities and their respective landscapes.[1] This absence, as we know from the voluminous work on imperial liberalism, has historically been enabled by classificatory systems cultivated during the early modern period—what has been termed the "white geology"—according to which such taxa as colonized laborers would be rendered nonhuman and thus fungible, without agency or political consequence.[2] As such, the laboring nonhuman (e.g., the colonial subject, the indentured worker, or the miner) living in the "extractive zone" of occupied Burma is virtually invisible—off-sited both literally and imaginatively.[3]

In this chapter, I examine the imbricated phenomena of extractive capitalism, colonialist improvement projects, and the violent abstractions that gave rise to indentured labor and plantation monocultures through a reading of Amitav Ghosh's *The Glass Palace.* I interpret the novel as a documentary of an emergent system of "cheap nature" in colonial-era Burma—that is, the commodification of timber, petroleum, and peasant labor—as well as an oil fiction in its attention to the genealogy of the modern petro-state.[4] As I further demonstrate, *The Glass Palace* stages a materialist intervention that makes evident the immiscible histories of both: as "dead abstraction," petrol—the commodity form produced from petroleum extraction—is indistinguishable from timber and also labor.[5]

In tracing the historical confluence of emergent extractive economies and seventeenth- and eighteenth-century theories of ontology qua taxonomy through Ghosh's fictional plantation economies, I likewise read the novel as a potent critique of imperial liberalism—in part engendered by René Descartes's formulation of the thinking subject (nay "Human") and emboldened in the era that follows, in the work of John Locke and, perhaps especially, John Stuart Mill.[6] In this way, I join Ghosh and others in emphasizing the role of race and associated forms of taxonomic violence in the foundations of liberal thought—a critical intervention that seems ever more urgent as contemporary critiques of development (and capital *tout court*) persistently reproduce the sorts of "simple" divisions, if also disingenuous indictments, that we see in the chapter-opening epigraph from George Orwell's *Burmese Days.*[7]

Despite (or perhaps because of) its resonance in thinking about new forms of development sanctioned by imperial interests, as a conventional/liberal critique of empire, *Burmese Days* is in many ways an unremarkable novel. Anticipating works such as V. S. Naipaul's 1979 *A Bend in the River,* which is similarly marked by primitivist caricature and a conspicuously imperialist prospect, it is a typical portrait of colonial racism in the guise of political critique. I cite it here because it concisely instantiates the liberal worldview that Ghosh takes to task; and its rendering of colonial-era Burma—rife with offensively physiognomic, often grossly abstract images of local residents—offers a perfect foil for *The Glass Palace.* Not to mention that if we consider the earlier passage in light of the increased attention to colonial-era improvement schemes reliant on the accumulation of fossil capital, the novel offers a useful frame for thinking about the development ethos implicit in Flory's remarks, dependent as they are on a neatly Manichean worldview that understands British might in contradistinction to an "unfortunate" peasantry.

Operating under the Lockean mandate to improve, through agricultural enclosure via "rational" modes of husbandry, development as such depended on a rigid distinction between colonial "society" and the uncultivated, or "rude," peasantry under its control.[8] Preceded by the partitioning of the reasoning modern subject (Human) defined over and against all nonhuman matter, local peasants—Burmese farmers, for example, who had previously worked on preindustrial plantations under local landlords—would be subjected to colonial-era taxonomies that posited them within a capacious category of unimproved Nature.[9] New classificatory systems, or taxonomies, imagined by Carl Linnaeus and others in the eighteenth century—who were themselves influenced by an early modern materialism that, *pace* Descartes, understood the natural world in terms of domination and subjugation—made possible imperial networks of trade reliant on such agricultural innovations as plantation monocultures as well as on new forms of mechanized labor: "Nature's activity [would become] a set of things [that could be] yanked into processes of exchange and profit, denominated and controlled."[10]

Now classified as commodities, workers could be thrust into a mode of production reliant on the reduction of complex ecosystems "into impoverished but exchangeable forms":[11] "ecological simplifications in which living things [were] transformed into resources [and] future assets by removing them from their lifeworlds."[12] Impoverished indeed, Orwell's characterization of Burmese peasant farmers is redolent with a particularly egregious form of onto-epistemological violence—a mode of violence both central to the so-called Cartesian revolution, or the "cosmology of the moderns," and essential to the ethically bankrupt, if also historically deficient, formulations of an Anthropocene that deny the long history of uneven development characteristic of global capitalism.[13] As is also demonstrated in Ghosh's resolutely materialist depiction of occupied Burma, the presumed separation of the knowing subject and the observable object, which forms the basis of the colonial episteme, was fundamental to contemporaneous modes of production as well as to theories of being that continue to infect popular thinking about Human and nonhuman communities—"Human" here understood as a restrictive category denoting a small segment of the global population and thus a vexed referent for a particular type against which the category of "nonhuman" is made possible.[14]

In an effort to move beyond such simplifications, I read *The Glass Palace* against the sorts of historical abstractions at work in independence-era novels such as Orwell's and in more recent instantiations of imperial liberalism such as we see in discussions around the Anthropocene—a stratigraphic term designating an age of humans and coined to explain anthropogenic climate change. Ghosh presents us with a historical epic that traces several generations of the family of its principal character, Rajkumar, across Burma, India, and Malaya—all British colonies in the century or so that the novel spans. Rajkumar's tale begins with the siege of Burma's royal palace and the fall of Konbaung, the last dynasty, and his story effectively concludes in the aftermath of World War II. Characteristic of the author's broader oeuvre in its attention to environmental violence and the material impact of the colonial encounter, the novel presents a robust ecological history of colonial Myanmar in its representation of the nation's teak and rubber economies and the families whose lives are irrevocably transformed by colonial occupation.

Rather than eschew the evident substrate of the peasants' misfortune, the novel foregrounds the material means of their dispossession. So too does Ghosh meticulously illustrate the imaginative coupling of peasant laborers with commodity crops such as teakwood, which was both a product of the previously mentioned cosmology and a necessary taxonomic condition for successful plantation economies in Burma and elsewhere. Only as constituent parts of such a system could the otherwise variegated category of "unimproved" nature—Human and nonhuman—be subordinated to a universalizing scalability so critical to the domestication of native teak forests, "teak being ruled, despite the wildness of its terrain, by imperial stricture in every tiny detail."[15] Like the teakwood qua timber, peasant farmers would also be "ruled . . . by imperial stricture."[16] Farmers such as Rajkumar were thrust into a category of labor wherein they were effectively denied autonomy and rendered so many "dead abstractions" indeed—concise figurations, that is, of "cheap nature," which is herein understood as "capitalism's ontological praxis."[17]

Put more simply, cheap nature refers to the cheapening of work in which the labor of "many humans—but also of animals, soils, forests, and all manner of extra-human nature—[is rendered] invisible or nearly so."[18] Jason Moore explains such modes of cheapening as the process through which capitalism must "[convert] the living, multi-species connections of humanity-in-nature and the web of life into dead abstractions."[19] Marx would characterize this process similarly, although, significantly, he replaced "humanity-in-nature" with humanity *and* nature. Thus, the production of a sort of cheap nature/labor in *Capital* is rendered as "a degraded and almost servile condition of the mass of the people . . . and the transformation of their means of labour into capital."[20] The dissolution of "the unity of living and active humanity with the natural, inorganic conditions of their metabolic exchange with nature, and hence their appropriation of nature," which Marx also outlines in *Grundrisse*, similarly privileges such a rift; for Ghosh, though, this relationship is somewhat more complicated.[21]

Following Moore's understanding of cheap nature, then, I read *The Glass Palace* as in fact a documentary of the cheapening of Myanmar's colonial landscape through the domestication of its native forests and the commodification of the local peasantry—"documentary," that is, insofar as Ghosh records, with vivid imagery, Burma's denuded forests. Cheapening is here understood as "a set of strategies to control a wider web of life [and] . . . a [form of] violence that mobilizes all kinds of work—human and nonhuman, botanical and geological."[22] As Rajkumar remarks, "It wasn't that you were made into an animal. . . . [N]o, for even animals had the autonomy of their instincts. It was being made into a machine: having your mind taken away and replaced by a clockwork mechanism."[23] Here the less-than-human subject is made machinelike by the mandate of plantation logic: "New patterns of work were being invented [which relied on] new patterns of movement, new ways of thought."[24] That is, "capitalism's first great remaking of planetary life . . . was scarcely possible without a revolution in ways of thinking and seeing the world."[25]

Significantly too, and as I have suggested earlier, this process of cheapening, which Ghosh illustrates primarily through teak, was predicated on the same epistemological and material foundations that allowed for Burmah Shell's speculative ventures in the latter part of the nineteenth century. The aforementioned classificatory systems made possible the commodity forms of teak and petrol that formed the basis of the colonial economy and emboldened imperial interests well after independence. The oil industry figures as an uncannily absent presence, enjoying only a few errant references, "so contained within the quotidian landscape of modernity that it does not present itself to view," but it haunts the novel's foreground.[26] Thus, while the reader might muse, à la Rajkumar, "A war on wood? Who's ever heard of such a thing?," the bombing of the oil fields toward the close of the novel will come as no surprise.[27] As we also come to learn, the Burmese petro-state was well established in the eighteenth century, inviting British interest in "earth oil" decades before the first Anglo-Burmese war in 1824—that is, prior to the consolidation of imperial Britain's Burmese colony.[28] Oil speculation would ultimately rival interest in teak, while both forms of energy were co-constitutive of the colonial economy and the formation of the modern state.

Although the novel has rarely been read as an oil fiction, with most critics taking on what seems the more overt critique of empire and thus effectively severing fossil capital from occupied Burma's political economy, *The Glass Palace* offers a remarkable glimpse into the complicated genealogy of the nation as a modern petro-state.[29] British Petroleum currently owns Myanmar's vast oil reserves, but it was their predecessors, first the Burmah Oil Company and later Standard Oil and Burmah Shell, that would help to transform the nation's plantation economy into a modern extractivist regime by co-opting a centuries-old tradition of oil extraction that had been effectively nationalized by the Burmese monarchy. Thus, after tracing the emergence of cheap nature in nineteenth-century Burma through the nation's teak economy, I offer a petro-aesthetic critique that makes visible the intersections between the aforementioned colonial-era taxonomies and the "thick ooze" percolating in the hillsides beyond the colonial capital at Rangoon. In this way, I place Ghosh's novel within the generic category of "oil fiction" that petro-critics such as Imre Szeman and Graeme Macdonald have increasingly embraced as a means of building on the narrower framework of Ghosh's earlier coinage of "petrofiction."[30]

I also unsettle the neat temporalities attendant to dominant theories of the Anthropocene, which foreclose the sorts of radical imaginings that might allow for one to read *The Glass Palace* as anything other than a chronicle of the teak industry. As oil fiction, *The Glass Palace* provides a possible model for imagining what critics have lately termed a "Thermocene"—a geological epoch forged through carbon emissions, whether wood or petrol—and thus a more productive heuristic for mapping the planetary transformations that mark the latter Holocene. As such, the novel forces a more nuanced engagement with critiques of anthropogenic climate change that seem insufficiently concerned with the material legacies of empire. Before offering a reading of the novel as a documentary of cheap nature, I thus turn briefly to its role in figuring what we might call a Thermocene aesthetics, if only to dispel materially deficient interpretations of our geological history. In this brief excursus, I provide the logical conditions for understanding the fluidity of the carbon economy in the ages of wood and oil and thus the necessity of reading *The Glass Palace* as an oil fiction.[31]

Anthropocene, Capitalocene, Thermocene

The novel begins in 1885—the year British Burma would become an autonomous province given sanction by then secretary of state Randolph Churchill and the year before Burmah Oil would establish its monopoly; the opening scene features the expulsion of the extant monarchy. As Ghosh's character Matthew points out, 1885 also marks another momentous birth: that of the "'motorwagon' . . . the small internal-combustion engine, the vertical crankshaft, the horizontal flywheel. . . . It had been unveiled that very year . . . in Germany by Karl Benz."[32] But it is teak, not petrol, that occupies the central frame. The timber economy is one in a series of imperial projects, each of which depends on the abstraction-cum-cheapening of the human, nonhuman, and more-than-human landscape.

Ghosh is careful to demonstrate the intersections between what are often read as succeeding generational investments in energy forms: illustrating a "proto-fossil fuel" economy, he invites the reader first into the teak plantations where competing energy technologies participate in the disruption of local economies and the devastation of local environments.[33] Adding oil to teak, and later rubber to oil in the case of colonial-era Burma—because energy histories don't operate along neat breaks—*The Glass Palace* dramatizes how teak, oil, and rubber are co-constitutive of the nation's evolving petro-economy. As the engine thrums on a 1939 jet during the second half of the novel, its passengers deftly soar above the thriving rubber plantations that provide for their immense wealth and help to make modern flight possible.

Foregrounding what Christophe Bonneuil and Jean-Baptiste Fressoz have termed a "political history of carbon"[34]—a history in which "fossil fuels should, by their very definition, be understood as a social relation [because] no piece of coal or drop of oil has yet turned itself into fuel"[35]—the novel intervenes radically in the aforementioned Anthropocene narrative from which such material histories have been persistently omitted. Such a history is also central to Ghosh's own *The Great Derangement: Climate Change and the Unthinkable*, in which the novelist foregrounds multiple energy sources—including wood, oil, and steam—as well as the role of empire in his alternative history of the carbon economy.[36] If we think not in terms of transitions or stages but instead in terms of crises of energy and capital (as well as their material continuities), such histories enable a more rigorous understanding of fossil capital's uneven histories. *Pace* historian Dipesh Chakrabarty, "human civilization . . . did not begin on condition that, one day in his history, man would have to shift from wood to coal and from coal to petroleum and gas."[37]

The novel forces consideration of the lineaments of an ascendant petrosphere remarkable for its uneven impact on figures like Rajkumar, not to mention the scores of other workers who populate the novel. As a praxis for a sort of Thermocene thinking, this allows for a broadening of the imagination, for a consideration of emergent forms of cheap nature and cheap labor that predate, for example, the 1784 steam engine that all too often functions as a trope in Anthropocene studies. Citing the "Romantic period as the moment during which the capitalism that now covers the earth *began* to take effect," Timothy Morton, for example, eschews consideration of the epoch's *longue durée*, thus foreclosing a reading of capital and its pernicious effects on climate that takes empire into account.[38] But empire must be taken into account, and this is precisely what novels like *The Glass Palace* afford, as does Moore's reformulation (following sociologist Andreas Malm) of a Capitalocene—a comparatively more capacious framework for thinking through the environmental impact of global networks of accumulation—which similarly undermines efforts to provincialize capitalism by blurring any easy distinction between the historical processes of capitalism and colonialism.

As a work about empire and also in many ways an oil fiction, *The Glass Palace* takes on the necessary material conditions of petrol production as well as the always intertwined phenomena of extractivist regimes and the dispossession of the poor and marginalized. So, too, when read as an "oil fiction," this energy-intensive novel instantiates a robust generic intervention—not so much a historical correction as the sort of broadening at stake in Szeman's coinage of the term. The novel marshals new forms of storytelling—explicitly energy intensive while also agentially rich, severed from the always already singular *Anthropos*—to tell a new story, a "geostory" of sorts, in the form of an oil fiction.[39] In line with Donna Haraway's imperative to look for new modes of storytelling, the novel makes clear that "the arts of a world powered by horses and the labour of bodies cannot help but be different from the expanded and compressed time, space and power of . . . petromodernity."[40] Accordingly, as petrocultural production, *The Glass Palace* replaces, for example, the individualist (and anthropocentric) pathologies of the modern novel—the "art" of which was forged at a very different moment in our energy history—with a rich polyphony of voices.

As such, Ghosh's historical novel departs substantively from popular novels like Orwell's. A materialist intervention par excellence, the novel offers an intimate glimpse into colonial-era labor regimes too often elided in the Orwellian fictions that replace material history with liberal lament. Its commitment to the legibility of labor and the correlative production of cheap nature through the dispossession of local farmers contests the Naipaulian abstractions of an earlier generation of postcolonial fiction writing. Reading *The Glass Palace* alongside Orwell, whose reception among the academic left continues to place him in a sort of progressive elite, also makes clear Ghosh's alignment with projects such as Haraway's in which the Kantian bootstrap narratives that necessitated the exploitation of Native persons and colonial environments are exposed in all their material horror.

Of course, discussions around what Romantic critic Alan Vardy has called "aesthetic enclosure"[41] are not new; nor are critiques of the "postcolonial picturesque"[42] tradition evinced by the works of Anglophone writers like Orwell and Naipaul. "Postcolonial picturesque" is a term coined by Pablo Mukherjee to mark the adaptation of a violent aesthetic tradition wherein marginalized persons, from itinerant laborers to decommissioned soldiers, were rendered picturesque objects in mid-eighteenth-century landscape paintings. Regarding the erasure/objectification of labor, whether the colonial subject or the rural peasant (their violent abstraction such that farmers, for example, as well as forms of nonhuman labor are all but invisible), capitalism as a project and process aligns not only with such historically constituted modes of dispossession as we see in colonial Burma but also with aesthetic forms that have long hinged on such imaginative erasures.

In opposition to such problematic representations of the colonial landscape, *The Glass Palace* presents a sweeping historical epic that foregrounds labor and documents, in excruciating detail, the transformation—ecologically and thus economically—of colonial-era Burma wherein informal labor regimes from the subcontinent would find themselves on newly minted plantations owned by colonial landlords and their local proxies. Like his character Dinu—a photographer, and Rajkumar's son—Ghosh paints a landscape "replete with visual drama: the jungle, the mountain, the ruins, the thrusting vertical lines of the tree trunks juxtaposed against the sweeping horizontals of the distant sea."[43] But where Dinu "labored to cram all these elements into his frames," Ghosh undermines this imperialist prospect by lingering closer to the soil.[44] Here the novelist, unlike his character, resists the urge to imaginatively enclose the native landscape.

Documenting Cheap Nature

Although teak, rubber, and petroleum are different substances—the slippery technics of petroleum in particular marshaling an altogether different means of extraction and engendering new forms of precarity—each economy operates under the aforementioned principle of improvement, or cheapening in the onto-epistemological sense. On Ghosh's teak plantations "[the landscape] stretch[ed] away as far as the eye could see. . . . There were orderly rows of saplings, all of them exactly alike, all of them spaced with precise, geometrical regularity."[45] The central plantation in the novel—Morningside, which would produce first teak and later rubber—was, according to one of Ghosh's principal characters,

> like neither city nor farm nor forest; there was something eerie about its uniformity; about the fact that such sameness could be imposed upon a landscape of such natural exuberance. [Dolly] remembered how startled she'd been when the car crossed from the heady profusion of jungle into the ordered geometry of the plantation. "It's like stepping into a labyrinth," she said.[46]

Later descriptions of its internal workings reveal a more disturbing glimpse into the human and environmental costs of teak: "To look at it, it's all very green and beautiful—sort of like a forest. But actually it's a vast machine, made of wood and flesh. And at every turn, every little piece of this machine is resisting you, fighting you, waiting for you to give in"[47]; "it was when [Dinu] crossed back into the monochrome orderliness of the plantation that he felt himself to be passing into a territory of ruin, a defilement much more profound than temporal decay."[48]

Morningside is a concise instantiation of cheap nature—its forests impoverished by commodity abstraction, and its laborers, human and nonhuman (in the case of Rajkumar's elephant comrades), reducible to so much "wood and flesh." This depiction of the geometrical regularity of the forest also testifies to the critical role of agricultural scale—call it rational improvement or call it imperial violence—so vital to the cheapening of the local landscape. Additionally, the reference to "flesh" forces consideration of, and also animates, the otherwise invisible labor regimes conscripted to transform rubber trees into, for example, a viable commodity for the nascent automobile industry. In this sense, Ghosh's novel, more than simply dramatizing the domestication of native forest to the mandates of plantation logic, *makes flesh* an otherwise abstract instantiation of cheap nature, "as though life had been breathed in a wall of slate."[49]

His elephants and the majestic teakwood itself are also endowed with a peculiar, unfamiliar sort of agency. Perhaps the more radical intervention is the novel's attention to nonhuman work:

> [Rajkumar] could not resist the spectacle of watching elephants at work: once again he found himself marveling at the surefootedness with which they made their way through the narrow aisles, threading their great bodies between the timber stacks. There was something almost preternatural about the dexterity with which they curled their trunks around the logs.[50]

In yet another scene, an elephant laborer exacts revenge against a colonial officer—possibly an authorial indulgence, but a compelling one nonetheless. The novel carefully attends to the ways in which Burma's human and nonhuman laborers were effectively rendered machinelike in the interest of amassing capital. Depictions of similarly subdued local landscapes abound in the novel: "To bend the work of nature to your will; to make the trees of the earth useful to human beings—what could be more admirable, more exciting than this?"[51]

Such passages provide a model for conceptualizing a resource aesthetics—a means of undermining "the resource logic of capitalism [that] presupposes that resources have no aesthetic whatsoever" and thus refusing an "aesthetic that effortlessly normalizes the brute inputs" of human and nonhuman labor long invisible to consumers of what we may surely call the postcolonial picturesque, here in the form of a petrified amorphous peasantry.[52] The novel simply refuses to normalize, if also to abstract, "the brute inputs of energy" on which the petro-economy depends. As if to drive home the material impact of the new energy regime, in one particularly prescient scene the brutal killing of a local missionary is coupled with the presence of a shiny new automobile: "a motorcar—a gleaming, flat-topped vehicle with a rounded bonnet and glittering, twelve-spoked wheels."[53] This scene, set in 1914, also positions the gleaming spectacle alongside scores of dead Burmese and Indian soldiers.

The novel thus makes visible the material means of production and extraction as well as the relations of resources like oil and rubber to capital abstraction. If the automobile is reducible to a symbolic evocation of mobility and freedom, the novel is interested in materializing the means of its existence—an existence made possible by the combined forces of plantation logic, technological ingenuity, and empire. Not unlike the earth-oil in the hills beyond Morningside, "the [teak leaf] . . . came from a tree that had felled dynasties, caused invasions, created fortunes [and] brought a new way of life into being."[54]

Reading Oil: A Petro-aesthetic Critique

While on the surface *The Glass Palace* is generally a story about teak and rubber, it is surely also an oil fiction. Its almost fleeting reference to oil is immaterial when one considers the evolution of Myanmar's carbon economy, not to mention the emergent automobile industry for which rubber would be just as critical: "[Rubber] was the material of the coming age; the next generation of machines could not be made to work without this indispensable absorber of friction. The newest motor cars had dozens of rubber parts; the markets were potentially bottomless, the profits beyond imagining."[55] It was also a source of colonial violence, as the reader learns that "every rubber tree in Malaya was paid for with an Indian life."[56] Unlike the British in the nation, "in Malaya the only people who lived in abject, grinding poverty were plantation labourers."[57]

British Malaya was critical to an imperial petro-imaginary, which was codified through the 1934 Petroleum Act. This act essentially consolidated British rule over regional petroleum reserves, which, as we also see in the novel, had been the province of local speculators for some time. Given its ascending hegemony then, the novel's oblique treatment of Burmese petroleum seems to attest to Ghosh's own assertion regarding the narrative illegibility of petroleum—a substance that seems to have resisted more explicit representation in the abovementioned modern "arts." Although "it does very little to point to the presence of fossil fuels in fiction, to go searching about for those few places where coal, gas or oil might resurface, receive mention or be extracted from the narrative. What we need instead is a new critical sensibility in our analyses of world literature."[58] We also need a new critical sensibility in our understanding of the origins of the carbon economy. And so despite the fleeting presence of "earth oil" in *The Glass Palace*, I read the novel as an oil fiction—a variant of Ghosh's much-remarked-upon category of petrofiction in which the trope of the oil encounter is sublimated within the broader lineaments of our contemporary petrosphere.

To approach literary productions in the age of oil from within the narrow frame of the conventional oil encounter—that is, "the intertwining of the fates of North Americans and those living in the Middle East around this commodity"[59]—is to deny what we ought rightfully understand as the mise-en-scène of all artistic production within the era of petromodernity. As Graeme Macdonald and others have argued, we need to understand oil fiction as more than a literal index of the relationship between energy and the arts. Particular forms of energy make ways of being in the world possible. That is, energy, the freedoms that it promises, and the cultural productions that it enables cannot be severed from one another. In addition, Ghosh's formulation of "petrofiction," depending as it does on the oil encounter, not only forecloses the inclusion of novels such as *The Glass Palace* but also denies entry to a novel like *Moby Dick*, which Macdonald suggests was a "prototype representation of a process endemic to the global history of oil extraction and petrochemical commerce."[60]

There is, nonetheless, explicit mention of petroleum in the novel. Beyond the aforementioned reference to burning oil fields, there is a long passage about Yenangyaung, where the "foul-smelling" stuff oozed endlessly to the surface:

> On the eastern bank of the river, there appeared a range of low, foul-smelling mounds. These hillocks were covered in a thick ooze, a substance that would sometimes ignite spontaneously in the heat of the sun, sending streams of fires into the river. Often at night small, wavering flames could be seen in the distance, carpeting the slopes. To the people of the area this ooze was known as earth-oil: it was a dark, shimmering green, the colour of bluebottles' wings. It seeped from the rocks like sweat, gathering in shiny green-filmed pools. In places, the puddles joined together to form creeks and rivulets, an oleaginous delta that fanned out along the shores. So

> strong was the odour of this oil that it carried all the way across the Irrawaddy: boatmen would swing aside when they floated past these slopes, this place-of-stinking creeks—Yenangyaung. This was one of the few places in the world where petroleum seeped naturally to the surface of the earth. . . . The gathering of the oil was the work of a community endemic to those burning hills, a group of people known as *twin-zas*, a tight-knit, secretive bunch of outcasts, runaways and foreigners. Over generations *twin-za* families attached themselves to individual springs and pools, gathering the oil in buckets and basins, and ferrying it to nearby towns. . . . Some of these wells were so heavily worked that they looked like small volcanoes, with steep, conical slopes. At these depths the oil could no longer be collected simply by dipping a weighted bucket: *twin-zas* were lowered in, on ropes, holding their breath like pearl divers. . . . Wooden obelisks began to rise on the hillocks, cage-like pyramids inside which huge mechanical beaks hammered ceaselessly on the earth.[61]

Perhaps the "shimmering" earth-oil labors too—bubbling to the surface and combusting spontaneously, as it were—thus contesting the aforementioned perception of its "brute" nature.

Reading the preceding passage, set a century ago, one might be forgiven the immediate visions of contemporary Myanmar, in which displaced citizens flock to these nearly exhausted oil fields. In Nga Naung Mone, the site of the nation's largest unregulated oil field, a small city has emerged wherein migrants have constructed dangerously shoddy derricks in the hope of earning a daily wage from the nation's refineries.[62] As in the case of nineteenth-century Burma, their labor will not sustain them; their (largely) unpaid work merely greases the much larger engine of British Petroleum or Chevron. The latter also owns a significant stake in the nation's oil industry.

That the current administration has opened the fields to general speculation speaks also to the persistent power of rhetorical abstraction: oil speculation continues to promise forms of freedom that have never been available to the nation's rural majority. Since independence in 1947 and then a brutal civil war, the shadow of empire persists in the form of a British stronghold on oil and precious minerals—all mined by a rural precariat in unspeakable conditions—as well as public-private partnerships between the government and multinationals like Chevron, which continues to tout the false promise of "economic growth."[63] In line with Ghosh's argument in *The Great Derangement,* such promises are but "grotesque fictions"[64]—evidence not only of the necessarily uneven forms of development that enable capital accumulation in the colonial and neoliberal eras but also of what we might call uneven and combined enlightenment.

Of course, however grotesque the promise of freedom or mobility, the shimmering stuff continues to represent an oasis of capital for a resource-starved population. And despite American and British stakeholders owning the lion's share of the nation's oil reserves—Chevron's most recent endeavor secured 2.6 million acres in the offshore Rakhine Basin, adjacent to the Rakhine State, where the Rohingya population live under what Amnesty International has characterized as apartheid—and knowing the abject conditions associated with petroleum extraction for industry workers, impoverished Burmese citizens remain undeterred. As in the example of the rubber plantations cited earlier, while American and British elites profit from the industry, laborers may "[live] in abject, grinding poverty," but they continue to show up daily for lack of an alternative.

In another example of the characteristic precarity of such labor regimes, Ghosh's fictional Yenangyaung is mirrored by the palm oil plantations that would eventually replace teak (and rubber in some instances) and, *pace* our discussion of faulty historical periodizations, persist in the putative age of petroleum—plantations like Morningside also present a resonant portrait of extractivist violence, one that might recall similar scenes in the Niger Delta. Described as having "air . . . the texture of grease,"[65] Morningside—at this point producing palm oil—is also a harbinger of Myanmar's current palm oil industry, which continues to decimate local village economies through illegal land grabs, resulting in an internal diaspora of the sort that we see in *The Glass Palace*.[66]

Returning to Yenangyaung, however, in our consideration of a novel that rightfully ought to be read as "oil fiction," it must also be noted that this "foul-smelling" place is especially significant for another reason: it was in colonial Burma in 1859 (and not Titusville, Pennsylvania) that the modern oil industry would be born. As the author states elsewhere,

> It could be said that the first steps toward the creation of the modern oil industry were actually taken in Burma. But where these steps might have led we do not know because Burma's attempts to control its oil came to an abrupt end in 1885, when the British invaded and annexed the remnants of the Konbaung realms, deposing

> Thibaw, the dynasty's last king. After that, the oil fields of Yenangyaung passed into British control, and, in time, they became the nucleus of the megacorporation that was known until the 1960s as Burmah Shell.[67]

Thus, while Ghosh chooses (in *The Glass Palace*) to sublimate the imperium of the oil industry, opting instead to conjure the grander sphere of influence that it would wield, the novel closely follows its ascendancy, opening with the siege of the Burmese palace in 1886. This is no accident. The history of carbon, for Ghosh, is a central determining factor of the loose constellation of discourses and material conditions that we call modernity, or more appropriately, modernities. Logically, then, the twin births of British Petroleum and a consolidated Indo-Burmese state constitute the temporal anchor in this otherwise sprawling historical novel, thus placing *The Glass Palace* alongside other works of oil fiction that similarly recognize the role of energy in any plausible narrative of modernity.

Conclusion

As an oil fiction, *The Glass Palace* gestures to the *longue durée* of carbon by opting to begin with wood and ultimately demonstrating that the "machine made of wood and flesh" may as well be the machine dripping with oil and blood. In documenting the ascendancy and imperium of colonial-era taxonomies that made new forms of cheap nature possible, the novel likewise speaks to the historical-material conditions necessary for the modern petro-state and thus the contemporary petrosphere. If the plantation, like its latter-day correlate the oil field, is reducible to a symbolic evocation of improvement in the Lockean sense (and certainly development in its liberal and neoliberal senses), the novel instead marshals a materialist critique to indict the sorts of imperial liberalism at work in novels such as *Burmese Days*. So, too, in undermining the facile time scales of Anthropocene narratives that begin with steam, *The Glass Palace* makes clear the imperialist origins of the latter Holocene, which should really be located centuries before the steam engine and surely also before the "satanic mills" of more popular Anthropocene narratives—those icons of *the* industrial age from which William Wordsworth would notoriously flee.[68]

1. William E. Connelly, *Facing the Planetary: Entangled Humanism and the Politics of Swarming* (Durham: Duke University Press, 2017), 34.

2. See, for example, Lisa Lowe's *The Intimacies of Four Continents* (Durham: Duke University Press, 2014). For a discussion of "white geology," see Kathryn Yusoff, *A Billion Black Anthropocenes or None* (Minneapolis: University of Minnesota Press, 2019).

3. See Macarena Gómez-Barris's *The Extractive Zone: Social Ecologies and Decolonial Perspectives* (Durham: Duke University Press, 2017).

4. Jason W. Moore, "The Capitalocene, Part I: On the Nature and Origins of Our Ecological Crisis," *Journal of Peasant Studies* 44, no. 3 (2017): 594–630.

5. Ibid., 598.

6. See Descartes's configuration of the subject in his 1641 *Meditations on First Philosophy* as well as discussions about nonhuman animals in his 1637 *Discourse on Method.*

7. In his correspondence with Dipesh Chakrabarty regarding the publication of Chakrabarty's *Provincializing Europe*, Ghosh remarks, "I do not understand . . . how it is possible to discuss J. S. Mill (or Bentham or any other 19th-century British liberal) without accounting for the place that the idea of race occupies in their discourse. . . . [T]o omit [race] is merely to ignore the ground on which liberal thought is built" (148).

8. See John Locke's 1689 *Second Treatise on Government.* See also John Stuart Mill's characterization of India's "rude" peasantry in *Considerations on Representative Government* (1861).

9. I use the term "preindustrial" despite the false periodization that wrongly privileges the 1784 pivot to steam as a "revolution." I use it because of its relationship to discussions of peasant labor within a Marxist view of political economy.

10. Raj Patel and Jason W. Moore, *A History of the World in Seven Cheap Things* (Berkeley: University of California Press, 2017), 47–48. See also John Bellamy Foster's discussion(s) of early-modern, Enlightenment-era, and ancient materialisms in *Marx's Ecology: Materialism and Nature* (New York: Monthly Review Press, 2000).

11. Ashley Dawson, *Extinction: A Radical History* (New York: O/R Books, 2016), 43.

12. Anna Lowenhaupt Tsing, *The Mushroom at the End of the World: On the Possibility of Life in Capitalist Ruins* (Princeton: Princeton University Press, 2015), 5.

13. Didier Debaise, *Nature as Event* (Durham: Duke University Press, 2017).

14. For discussions of onto-epistemological violence and the formation of modern categories of race (and of the human), see especially Sylvia Wynter's "Unsettling the Coloniality of Being/Power/Truth/Freedom: Towards the Human, After Man, Its Overrepresentation—An Argument." *The New Centennial Review* 3, no. 3 (2003): 257–337.

15. Amitav Ghosh, *The Glass Palace* (New York: Random House, 2002), 69.

16. Ibid.

17. Ibid.

18. Ibid., 600.

19. Ibid., 598.

20. Karl Marx, *Capital*, vol. 1 (London: Penguin Classics, 1990), 881.

21. Karl Marx, *Grundrisse* (New York: Vintage, 1973), 489.

22. Ibid., 3, 22.

23. Ghosh, *Glass Palace,* 522.

24. Ibid., 221.

25. Ibid., 605.

26. Imre Szeman and Maria Whiteman, "Oil Imaginaries: Critical Realism and the Oil Sands," *Imaginations* 3, no. 2 (2012): 55.

27. Ghosh, *Glass Palace,* 15.

28. Ibid., 122.

29. See, for example, R. K. Gupta "'That Which a Man Takes for Himself No One Can Deny Him": Amitav Ghosh's *The Glass Palace* and the Colonial Experience," *International Fiction Review* 33 (2006). 18–26.

30. Amitav Ghosh, "Petrofiction," in *Incendiary Circumstances: A Chronicle of the Turmoil of Our Times* (New York: Houghton Mifflin, 2005), 138–51.

31. Patricia Yaeger, "Literature in the Ages of Wood, Tallow, Coal, Whale Oil, Gasoline, Atomic Power, and Other Energy Sources," *PMLA* 126, no. 2 (2011): 305–10.

32. Ghosh, *Glass Palace*, 11.

33. Andreas Malm, *Fossil Capital: The Rise of Steam Power and the Roots of Global Warming* (New York: Verso, 2016).

34. Christophe Bonneuil and Jean-Baptiste Fressoz, *The Shock of the Anthropocene: The Earth, History and Us* (New York: Verso, 2016).

35. Andreas Malm, "The Origins of Fossil Capital: From Water to Steam in the British Cotton Industry," *Historical Materialism* 21, no. 1 (2013): 17.

36. Amitav Ghosh, *The Great Derangement: Climate Change and the Unthinkable* (Chicago: University of Chicago Press, 2016), 93, 108.

37. Ibid., 216.

38. Timothy Morton, *Hyperobjects* (Minneapolis: University of Minnesota Press, 2013), 4 (emphasis mine).

39. Donna Haraway, *Staying with the Trouble: Making Kin in the Chthulucene* (Durham: Duke University Press, 2016).

40. Imre Szeman, "Conjectures on World Energy Literature: Or, What Is Petroculture?," *Journal of Postcolonial Writing* 53, no. 3 (2017): 281.

41. Alan Vardy, *John Clare: Politics and Poetry* (New York: Palgrave, 2003).

42. Upamanyu Pablo Mukherjee, *Postcolonial Environments: Nature, Culture and the Contemporary Indian Novel in English* (London: Palgrave Macmillan, 2010).

43. Ghosh, *Glass Palace*, 349.

44. Ibid.

45. Ibid., 198.

46. Ibid., 199.

47. Ibid., 232.

48. Ibid., 336.

49. Ibid., 140.

50. Ibid., 458.

51. Ibid., 75.

52. Brent Ryan Bellamy, Michael O'Driscoll, and Mark Simpson, "Introduction: Toward a Theory of Resource Aesthetics," *Postmodern Culture* 26, no. 2 (2016): 2.

53. Ghosh, *Glass Palace*, 197.

54. Ibid., 71.

55. Ibid., 182.

56. Ibid., 233.

57. Ibid., 346.

58. Szeman, "Conjectures on World Energy Literature," 286.

59. Ibid., 282.

60. Graeme Macdonald, "The Resources of Fiction," *Reviews in Cultural Theory* 4, no. 2 (2013): 7.

61. Ibid., 122–23.

62. Adam Dean, "Drilling for a Dream in Myanmar," *New York Times*, May 22, 2017.

63. Per Chevron's company website, "For more than 20 years, Chevron, through its subsidiary Unocal Myanmar Offshore Co., Ltd., has worked with partners in Myanmar to spur economic growth and development."

64. Amitav Ghosh, "The Great Derangement: Fiction, History, and Politics in the Age of Global Warming" (Berlin Family Lectures, University of Chicago, Chicago, IL, October 7, 2015).

65. Ghosh, *Glass Palace*, 500.

66. Taylor Weidman, "The Human Cost of Palm Oil Production in Myanmar," *Al Jazeera*, January 4, 2017.

67. Ibid., 102.

68. William Blake, "Jerusalem," in *The Complete Poetry and Prose of William Blake* (Berkeley: University of California Press, 2008), 95.

8.

Aestheticizing Absurd Extraction

Petro-capitalism in Deepak Unnikrishnan's "In Mussafah Grew People"

Swaralipi Nandi

Named after the global oil giant, Patrick Chamoiseau's *Texaco* stages oil in a variety of forms—from the gigantic, menacingly standing oil tankers to the naming of the dusty shantytown built on the scraps of Texaco's urban center Fort-de-France to the characterization of the Urban Planner Christ as an insidious representative of petromodernity who aims to raze the habitation of the disenfranchised.[1] Yet, oil's insidious presence is felt most discernibly on the human lives it seeks to dismantle. Chamoiseau presents a *longue durée* of the modern Martinician state and its formation alongside an ascendant petromodernity of Martinique in which the age of colonial slavery flows seamlessly into the age of oil imperialism as a continuum of exploitation and dispossession. Thus, the human cost of oil invokes a trajectory of suffering and resistance that continues uninterrupted from the characters of Esternomme to Marie Sophie. Oil looms as a dark, omnipresent force that casts an eternal shadow as the chapters roll on from "Age of Crate Wood" to "Asbestos" and into "Age of Cement," documenting extraordinary human resilience against the onslaught of a petro-capitalistic urbanization.

The hagiographic nature of Chamoiseau's novel plays on ages, almost self-consciously commenting on their intersections and imbrications. Consequently, the connection, certainly not incidental, between slavery and the contemporary oil regime in Chamoiseau's novel proposes a potential critical paradigm for rendering oil visible through its depictions of the human costs of extraction. Petroleum commands an extractive ecology of its own, spawning a capitalist system that thrives on multifarious forms of labor exploitation beyond those at the site of production. Like other extractive economies such as mining and plantation agriculture, oil's transnational presence not only invokes a complex web of production and consumption affecting a vast geopolitical grid of military power but also spans across a global network of labor, usually precarious, whose routes are often drawn along imperial histories of human migration. This petro-matrix thus encompasses an immense swathe of transnational labor that is drawn on the grids of an international division of labor, concealed at the site of production, and embodies a "commodity fetishism"[2] that makes invisible the human cost of a petro-economy.

In this chapter, I work under the premise—to reiterate what we state in our introduction—of an understanding of oil fiction that takes into account the material conditions of its production in the context of extractivist regimes that thrive on the dispossession of the poor and marginalized. Asserting that petro-economies essentially replicate slave economies through overt parallels of borderless labor extrication from the global fringes, I look at how such energy slaves have been rendered visible, albeit through an aesthetics of absurdism, in Deepak Unnikrishnan's short story "In Mussafah Grew People."[3] I read Unnikrishnan's story on pod laborers by using Michael Lowy's concept of "critical irrealism,"[4] such that the aesthetics of "absurdity" of Unnikrishnan's vegetized humans is a potent critique of the capitalistic fantasy of an inexhaustible expropriation that thrives on the incessant supply of an exploitable environment and disposable labor and that endorses Andreas Malm's theory of fossil fuels as "the general lever for surplus value production."[5] Subsequently, I also invoke Raj Patel and Jason W. Moore's notion of "cheapness" to read the pod laborers as "a strategy, a practice, a violence that mobilizes all kind of work—human and animal, botanical and geological—with as little compensation as possible."[6]

On Oil Visibility: Where Should We Be Looking?

While oil emerges ubiquitous as a presupposed energy form fueling the motion of a global capitalist society, the "order" of its functioning has often been tagged as largely invisible, effecting an alienation of the commodity from its methods of production. Examining the visible markers of an oil culture—the train of pipelines—Graeme Macdonald argues that the "invisibility" of the global petro-system is "both culturally naturalized and internationally relative—structurally managed in the interests of an oil-reliant capitalism seeking to extend and perpetuate supply while downplaying the ongoing, exploitative shame of extraction and land dispossession and the inevitable endpoints of burning."[7] The question of invisibility that, as Macdonald explains, has to do less with a literal lack of vision and more with the "public indiscernibility" of oil's infrastructure and procurement, or in Imre Szeman's words, an "epistemic inability or unwillingness to name our energy ontologies,"[8] essentially banks on the unequal power relations of an oil economy. Macdonald's resounding questions on modes of "humanizing oil" and on forms of seeing and representing that can contribute to "meaningful forms of action"[9] echo the quest for a more tangible, praxis-based critical paradigm of approaching oil through cultural texts.

Overall, there has been a consensus among oil critics about oil's determinate invisibility, so the issue has subsequently been approached in two ways. First propounded by Amitav Ghosh in his indictment of the absence of the great American oil novel, the lack of visibility has ever since met numerous rejoinders proclaiming the "presence" of oil in various past fictions.[10] Graeme Macdonald and Rosalind Williams both cite *Moby Dick* as one of the earliest oil novels that engages with the human quest for and consumption of oil, offering valuable insights into energy's embeddedness in human experiences. While Macdonald goes on to include *The Great Gatsby* and *The Corrections* in the list of oil fictions,[11] Williams digs into *Country of Pointed Firs* to explore how it documents energy's retreat and into the *Education of Henry Adams*, which features force, in the form of a dynamo, as a clue to energy's intricate connection with human history.[12] The thrust of such readings is based on energy's inevitable ubiquity in expressions of the human experience—as Macdonald succinctly proclaims:

> By this reckoning, is not all fiction from, say, *The Great Gatsby* (1925) to *The Corrections* (2001) "oil" fiction? Is not any work by John Steinbeck or William Faulkner or Richard Wright, Philip Roth, or David Foster Wallace petrofiction? All modern writing is premised on both the promise and the hidden costs and benefits of hydrocarbon culture.[13]

Macdonald posits the classic contradiction of oil's cultural imaginary of being "everywhere in literature yet nowhere refined enough—yet—to be brought to the surface of every text." The onus thus falls on the "new generation of oil-aware petro-critics," who must embark on extracting a petro-presence that sits "untapped, bubbling under the surface, ready to be extracted."[14] Given the paucity of literary texts that address oil up front, "oil awareness," Szeman says, further explicating Macdonald's conjecture, thus entails text mining of past and contemporary literary archives to locate oil beyond where it figures explicitly, delving deep into oil's symbolic import embodied in "dispositions, expectations, capacities and desires that energy excess makes possible."[15]

Another way to engage with petroleum as more than a psychological abstraction of speed and fluidity of motion or cultural "expectations and desires of the modern world" entails plunging deep into the materiality of the murky world of petroleum as a commodity in the global world-system. The oeuvre of postcolonial literature on a Manichean oil economy that Macdonald mentions—Abdelrahman Munif's *Cities of Salt*, tracing the genealogy of oil in the Saudi desert; Ralph de Boissière's *Crown Jewel*, which takes up workers' strikes in Trinidadian refineries; George Mackay Brown's *Greenvoe*, highlighting the North Sea oil strike; Carlos Fuentes's *The Hydra Head* and Patrick Chamoiseau's *Texaco*, depicting the local resistance against corporate oil; and Nawal El-Saadawi's *Love in the Kingdom of Oil*—attests to powerful articulations of oil encounters in world literature.[16] Szeman explains it as postcolonial literature's trajectory of engaging with the "lived" experience of global imperialist forces, asserting that postcolonial writers like Munif can "grapple with the social power of oil" because the "significance of energy in shaping modernity is more readily apparent to those living in spaces where petroculture has yet to take hold,"[17] just as the colonial system was more evident to those who endured it. Adding Chris Abani's *Graceland* and Helon Habila's *Oil on Water* to the list embodies a considerable shift toward representing oil, visible on the surface as a tangible, debilitating force interminably disrupting human lives on the periphery.

Macdonald thus identifies the typical oil novel as one with the recurrent presence of "a sinister corporate interloper, working behind the scenes with local elites to change the destiny of local lives by dragging them and their territory into petromodernity."[18] Indeed, in all the texts just mentioned, oil comes to people in the form of a colonial intruder, violently usurping local lands through its corporate commissaries. Yet, what Macdonald's list does not represent, though he summarily emphasizes oil's essential transnational disposition, is its massive effect beyond the zones of contact, fueling myriad other forms of violence and human perpetration. Oil spills over in the form of petrodollars, spawning an immense network of global labor migration in oil economies so utterly deterritorialized that all notions of "local" are dismantled. The huge swathes of undocumented, unskilled laborers migrating from the margins of global economies to the UAE, Oman, Kuwait, and the other Gulf nations and sustaining the extravagant megacities of the petro-empires of the Gulf testify to a system of utter exploitation, inhuman extraction, and human rights violations. Critiquing oil thus entails including in the discourse its concrete human cost and its apparent immutability, as Brent Ryan Bellamy and Jeff Diamanti assert:

> You cannot see energy in the way that you can see a barrel of oil, because energy in the concrete is still abstract, and an energy system fueled by fossil fuels is more abstract still, even though it is determinate of virtually all economic and political capacities today. . . . The critique of energy is . . . a critique of the many barbarisms that flow from the contradictions of late fossil capital; and it is a critique of a fossil-fueled hostility to the very notion of social revolution—and hence of the very notion of structural dependence too.[19]

Mimicking the elusiveness of oil, this human workforce is largely invisible in both cultural and political imaginaries. They rarely appeared in canonical Anglophone postcolonial fiction until recently—barring maybe during the brief oil encounter of Alu in the mythic state of Al-Ghazira in Amitav Ghosh's *Circle of Reason*—which explains their indiscernibility in the context of the unequal terrain against which literary texts are globally disseminated and consumed. Benyamin's runaway hit *Aadujeevitham* (translated as *Goat Days*), which poignantly narrates the story of an abused Indian migrant worker in Saudi Arabia, brings the Gulf's undocumented labor force into wider literary discussions. The last couple of years have seen a host of other South Asian Anglophone fictions focusing on this group—Shyamlal Puri's *Dubai Dreams* and *Dubai on Wheels* on the undocumented taxi drivers of Dubai; Nikhil Ramteke's *365 Days* on the dark underbelly beneath Dubai's ultramodern urban shimmer; Mohanalakshmi Rajakumar's *Migrant Report*, presenting a motley band of migrants and their stories; Saud al-Sanaussi's *The Bamboo Trunk*; and Deepak Unnikrishnan's *Temporary People*, which rose to global prominence following its Man Booker Prize. These literary texts counter what Victor Turner would call the "structural invisibility" of the expatriate "liminal personae" in the cultural texts—the state of being at once "no longer classified and not yet classified."[20] Gulf migrant labor is absorbed into the fabric of capitalistic machinery and yet is not recognized as the legal labor-citizen, and so the precarious conditions of this labor—as I argue in the following section—replicate those of slavery, suggesting a connection between fossil fuel economies and the system of slavery that goes beyond the ethical parallels suggested by Jean-François Mouhot.[21] I read Unnikrishnan's "In Mussafah Grew People" as a potent mode of making the energy slavery visible as well as an exploration of aesthetic modes on exploitation beyond genres of realism.

From Petromodernity to Petro-labor

Published in 2017, *Temporary People*, by the Abu Dhabi–based Indian writer Deepak Unnikrishnan (he is also a resident of the United States), documents in utterly disjointed, fractured narratives the innumerable "temporary" lives of the undocumented Gulf migrants, mostly Malayalees who are natives of the Indian state of Kerala. Unnikrishnan weaves a rich polyphony of diverse labor tales through bizarre stories—Malayalee workers are grown in pods in a scientific project; laborers fall off rooftops and their pieces are taped together; laborers swallow passports and turn into passports for easy passage; a red phone teleports; an elevator molests children; cockroaches walk on two feet and speak Malayalam, Arabic, Farsi, and Tamil, among other languages; and tongues break free from young boys' mouths, releasing their words all over the street. Invoking literary masters of absurdism such as Kafka, Kundera, and Rushdie—whom Unnikrishnan claims are his inspiration—he documents with all this surreal shapeshifting a mélange of dystopic migrant lives without the usual polemics of testimony narratives. The polyphony of these diverse voices is complemented by a splintering of literary forms in which fiction and poetry, news reports and epics, blend seamlessly into one another, offering a complex collage of unsettling images that, as reviewer Vani Saraswathi asserts, doesn't overtly attack the Emirates but "disembowels the shiny and the new, laying bare the ugly."[22]

Unnikrishnan's "In Mussafah Grew People," a story in *Temporary Lives*, is one of the most haunting tales. Sultan Mo-Mo runs a little sultanate named Mussafah that reels under Dubai's megalomaniac dreams of development. The sultan chances upon three Malayalee workers who reveal the secret behind Dubai's fanatic growth—that it grows its own labor. What follows is a fantastical tale of an agroproject led by Dr. Moosa, who has invented a technique for growing human laborers in pods. A series of failures leads to a variety of human crops of the right consistency—MALLUS, or the canned Malayalees—who have a life-span of twelve years before they can be disposed of. This stunningly profitable production of an unlimited supply of free labor is cut short when Dr. Moosa changes his intent, infusing his human "produce" with reason, minds difficult to tame, and a longer life-span. Dr. Moosa is arrested and executed, and the potentially dangerous human seeds are transported to the desert in crates to be burned in order to prevent any potential labor unrest. The story then moves on to the second part, in which rebel worker pods, which survived the extermination and were strategically smuggled out to create political unrest, are woven into a dystopic story of a rebel regime that is equally oppressive. Invoking the dystopic visions of George Saunders, as several reviewers observe, "In Mussafah Grew People" offers a potent critique of an oppressive labor regime that replicates the structures of slavery, portraying the oil economy as an essentially violent and dehumanizing force, the rot of which taints even a reactionary rebellion.

Not unlike other literary works on oil, "In Mussafah Grew People" does not feature oil explicitly, and yet Unnikrishnan cleverly weaves oil in as an emblematic but powerful, all-pervading presence. Abu Dhabi's phenomenal growth makes Sultan Mo-Mo's palace stink of petrol, the metaphorical driving force behind the sheen of development. A more direct reference to the power of oil as the magic elixir that drives the world is found in the story's allusion to Goran Bregović's boisterous tribute to gas. The carnivalesque mood of the original lyrics of Bregović's song—"Gas, gas, Gas, gas, GAS / Allo allo eh / Ritam ritam"—sarcastically proclaims petromodernity's great leveling power for the poor: "Even the Good Lord is happy when the poor have fun / . . . We'd better sell the little we have / And make a big party."[23] Yet Bregović's lyrics that suggest the fuel's power of instant gratification and the overall euphoria it induces—"Gas, to the dashboard / Let's go pretty girl / Push the pedal all the way / And on, As long as there's a gasoline"—take the shape of an ominous caricature when sung as a collective anthem by Rahmat, the sultan's truck driver, and, ironically, the Malayalee victims whom he is about to leave in the desert to die. The gas song is a gruesome reminder of petrol's power to erase latent violence in a euphoria of apparent agency. But the reference to Bregović's gas song is more than just a sarcastic tribute to the consumptive power of petromodernity. Bregović forms just a piece of the global mélange that is Mussafah, which houses UK-trained administrators, Sri Lankan doctors,

Malayalee laborers, Pakistani transport owners, and Afghan truck drivers, all in an inextricable labor relation to an economy powered by petrol. As the Gulf reaps commodities (Sultan MoMo lives on popular global consumables such as "fiery salsa," Guinness, "hash," and American soap opera songs) and labor from the diverse locales of the globe, the story reasserts the essential transnationality of the petro-economy that calls for a petrocultural perspective sufficiently capacious to encompass what Amitav Ghosh calls a dispersion of oil in a "bafflingly multilingual [space] . . . [,] a space that is no place at all, a world that is intrinsically displaced."[24] "In Mussafah" as an oil fiction thus invokes both a petro-critical paradigm that looks beyond oil as a commodity and a consequent petro-discourse that is built on the tropes of energy consciousness in the Global North and myopically tied to a logic of consumption.

As I have mentioned, a distinct objective of this chapter is to explore a petro-critical paradigm that engages more actively with the question of labor in an oil economy, a trope that has been substantially missing in the existing discourse on petrocultures. The assertion of Sheena Wilson, Adam Carlson, and Imre Szeman in their introduction to *Petrocultures: Oil, Politics, Culture* of "the newfound awareness of oil's importance to our sensibilities and social expectations" and their emphasis on "the recognition that over the course of our current century we will need to extract ourselves from our dependence on oil and make the transition to new energy sources and new ways of living"[25] are essentially entrenched in a consumerist perspective. It is thus that Touraj Atabaki, Elisabetta Bini, and Kaveh Ehsani, in their significant collection *Working for Oil: Comparative Social Histories of Labor in the Global Oil Industry*, vehemently critique the erasure of labor in contemporary oil studies, blaming it on the usurpation of the oil discourse by cultural studies. Launching a vehement attack on the field of Energy Humanities, they maintain that

> these petrocritics are most concerned with how systems of energy provision have shaped public culture and the collective imaginaries of everyday life, yet they show surprising[ly] little interest in the people who work, or have worked[,] in [the] oil and gas industry.[26]

Atabaki, Bini, and Ehsani take to task cultural/literary critics (e.g., Stephanie LeMenager), historians (e.g., Christopher F. Jones), and Marxist scholars alike for ignoring class relations in the oil complex and for not choosing as the focus of their investigation the "lived experiences of labor."[27] It is in this context that my reading proposes a significant call for a distinctly postcolonial petro-discourse that entails a more critical emphasis on extractible petro-labor from the Global South. Such a petro-discourse reconceptualizes the petro-economy as essentially an ideological system of labor relations shaped by, as Michael Rubenstein puts it, "profoundly uneven distribution of oil's benefits and consequences to peoples and territories around the globe."[28]

Labor and Energy Regimes

Indubitably, labor forms a fundamental component in understanding energy systems. The fossil fuel regime, as Andreas Malm convincingly argues, is essentially built on accessibility to exploitable labor. Asserting that the transition to coal power from waterpower was spurred not by its relative cheapness but by the availability of cheap labor that eased transport, Malm assigns materiality to energy transitions as an integral part of the proletariat's relation to the industrial capitalist system:

> Steam was a ticket to the town, where bountiful supplies of labour waited. The steam engine did not open up new stores of badly needed energy so much as it gave access to exploitable labour. Fuelled by coal instead of streams, it untied capital in space, an advantage large enough to outdo the continued abundance, cheapness, and technological superiority of water.[29]

Essential to Malm's theory of energy transition is thus the malleability of coal to suit capitalism's tenets of time and place abstraction that allowed for a more amorphous, more transitory source of energy that maximized control over labor.

Another prominent feature of the petro-economy of the Gulf is the easy exploitability of labor. Carl Skutsch estimates that the total number of migrants in the Gulf states grew from 2 million to 5.5 million between 1975 and 1985, after the 1970s oil boom,[30] and this phenomenal upsurge came with draconian labor systems such as the *kafala* that replicated conditions of trans-Atlantic slavery.[31] Entirely dependent on a local sponsor and deprived of citizenship rights, the *kafala* laborers—or petro-precariats,[32] as I henceforth call them—are inclusive of but not necessarily rig workers in direct communion with oil; rather they comprise the forgotten human actors slaving in the

Gulf system of capitalism, essentially bred and sustained by the power of oil. Explicit comparisons of the *kafala* system with slavery abound. Kurt Hauser calls it "a modern day chattel slavery, whereby the employer effectively owns the worker,"[33] while the Human Rights Watch lists *kafala* under contemporary modes of slavery. The Gulf's petro-economy thus invokes a relational investigation of energy modes and slavery that suggests a petro-critical paradigm beyond petroleum's commodity value, which accounts for its conceptualization as a system of labor extraction.

Existing scholarship on "energy slavery" seems inadequate to address the petro-precariats, however. While the connection between slavery and fossil fuel regimes has already been explored by scholars such as Andrew Nikiforuk (*The Energy of Slaves: Oil and the New Servitude*) and Jean-François Mouhot (*Energy Slaves*), both emphasize the metaphorical parallel between oil systems and slavery systems rather than focusing on the actual human actors in those systems. Their theoretical point of contention is again the logic of consumption, which parallels the contemporary energy consumer with a slaveholder. Nikiforuk compares the labor-saving machinery with "energy slaves" that make "ordinary people in geographies blessed with oil akin to slave masters, their powers rivalled only by the gods." He further explicates:

> Many North Americans and Europeans today enjoy lifestyles as extravagant as those of Caribbean plantation owners. Like slaveholders, we feel entitled to surplus energy and rationalize inequality, even barbarity, to get it. But endless growth is an illusion, and now that half of the world's oil has been burned, our energy slaves are becoming more expensive by the day.[34]

For Mouhot too, energy slavery is more of an ethical problem for the petroleum consumer of the Global North than for one of the Global South. The consumer from the North feels morally ambivalent about his energy consumption:

> Like him, I am a large slave-holder. Like him, I consider the idea of owning slaves to be abhorrent but feel as though I cannot really do without them. . . . Like him, I feel these slaves have a corrupting influence on me and on society in general.[35]

Both Nikiforuk and Mouhot personify the nonhuman fossil fuel–driven machines as energy slaves, while they are largely silent about human energy slaves, save as a passing reference. To Mouhot's credit, he does mention how energy slaves invisibilize human slavery elsewhere—for example, in China,

India, and Africa, where "workers labor from sunset to sundown harvesting cotton and cocoa beans and producing computer tablets for consumers in the rich world."[36] Yet such blatantly generalized and overtly decontextualized descriptions of slavery create more of a disjuncture than a correlation between it and petroleum. It is in this context that Unnikrishnan's story makes an important intervention, bringing together slavery and the petrodollar economy in a haunting tale of labor harvesting, even if he does so through a fantastical fiction.

In the tale, the clandestine agroproject ensures an incessant supply of labor in Dubai. Laborers are grown in flowerpots through a technique that was perfected after a series of failed experiments yielded partially formed humanoids and dwarfed human bodies. Finally, certain tweaks in the experiment—"adjusting the crop's exposure to light, giving the soil some air,"[37] using customized fertilizer, and keeping the environment climate-controlled—yielded the desired products that, after being picked, washed, and quality-checked, or "cerebrally customized," together made up an innocuous labor force named MALLUS (Malayalees Assembled Locally and Lovingly Under Supervision). The explicit reference to the Malayalees has been emphasized overtly, as the story offers their specific historical origin as well: this was "a certain kind of man, native to soil where thousands of years ago King Mahabali rule[d]. They call this man 'The Malayalee.'"[38] While such direct references, anomalous to the conventions of the allegorical form that Unnikrishnan tries to invoke in his tale, leave nothing to the reader's imagination about the laborers' ethnic identity, they facilitate the essential contextualization of the Gulf's labor migration in a specific history of world-system oil capitalism. Kerala's "Gulf boom," which saw a mass migration of sorts to the Gulf states following the oil boom, has been read as an economic phenomenon fostering domestic affluence through remittances and as a curious sociological case study of the migrant's return home. In several cultural texts, especially films, the Gulf *karan*, or the Gulf migrant, projects a cultural figure of affluence,[39] as in the *Diamond Necklace*; of alienation, as in *Varavelpu* and *Garshom*; or of humor, as in *Thinkalazhcha Nalla Diavasam* and *Akkare Ninnoru Maran*. The Gulf was thus the desired, exotic locale one would want to escape to, a magic elixir for the disgruntled life in the homeland. As Mohamed Shafeeq K points out,

> Gulf was a phantasmic presence for almost the entire 20th century Malayalam film. . . . Gulf thus became the symbol of not just instant success, [but] often of the

> outsider and at times of the undeserving, an American dream where one is surely and swiftly rewarded brooking no challenge from social relations back home, and even threatening to change the social equations at home.[40]

But more recent narratives of the Gulf depict bleaker visions of the migration. The journey to the Gulf has been portrayed as a perilous one, as in the recent film *Pathemari*, and the dhows of Beypore have served as a long-standing cultural metaphor for the excruciating, often illegal passage to the Middle East. Several other recent texts have portrayed the Gulf state as a dystopic space of inhuman exploitation for migrant labor, as in fictions like Benyamin's *Goat Days* (about a Malayalee migrant enslaved by an Arabic herdsman) or in commercial movies such as *Gadamma* (about the abuse of the female domestic-labor migrant) and *Arabikkatha* (about the migrant Malayalee workers in Dubai). Unnikrishnan's fiction adds to the list of narratives of the petro-precariats' horror in the Gulf but stands apart for its aesthetic choices. Moving beyond the just-mentioned texts, with their modes of realism and testimony narratives that typically recount the atrocities perpetrated on the Gulf migrants through intense, realistic portrayals of their suffering, Unnikrishnan's fantastical tale of the absurd extraction offers a caustic critique of the petro-labor regime that demands an incessant supply of dehumanized, slavelike labor. I read the absurdity of the tale through Lowy's concept of "critical irrealism," an aesthetic mode that not only signals an absence of reality but also retains the power to present social reality critically.

In introducing the term "critical irrealism" in the essay "Moonlit Enchanted Night," Michael Lowy poses it as a rereading of Georg Lukács' concept of "critical realism." Commenting on the long tradition of Marxist literary studies and particularly on Lukács' insistence on an aesthetic theory of critical realism as "the only acceptable form of art, and the only one that can have a critical edge in relation to contemporary social reality," Lowy assesses it as "exclusive and rigid."[41] As Lowy points out, Lukács' preference for the "literature of realism" as a "truthful reflection of reality" leads to his rejection of the most important modernist writers, such as Joyce, Kafka, Musil, Proust, Faulkner, and Woolf, among others. In contrast, Lowy argues—citing Hoffman, Kafka, and Huxley—for a more inclusive paradigm for representing reality:

> Could we not see in critical irrealism a complement of critical realism? By creating an imaginary world, consisting of fantastic, supernatural, nightmarish and *non-existent*

> forms, is it not able to illuminate aspects of reality in a way that clearly distinguishes it from the realist tradition?[42]

Lowy thus defines "critical irrealism" as an aesthetic mode that does not follow the rules of "the 'accurate representation of life as it really is,' but that [is] nevertheless critical of social reality."[43] This mode can be a potent expression of resistance against capitalism's ambiguous, multifaceted violence, for as Lowy asserts, "even when it takes the superficial form of flight from reality, critical irrealism can contain a powerful implicit negative critique, challenging the philistine bourgeois order."[44] According to Lowy, the "word 'critical' does not necessarily imply a rational approach, a systematic opposition or an explicit discourse" but rather invokes "features of protest, outrage, disgust, anxiety or angst"[45] that embody a potent critique of the capitalistic world order. Although Sourit Bhattacharya often uses "critical irrealism" interchangeably with the "marvellous real" or the more popular "magical realism," I find his impassioned argument about the distinction between the first and the third terms convincing. Bhattacharya makes a distinction between magical realism, the juxtaposition of postcolonial oral traditions with colonial European modernity in a form that has been sufficiently marketized in the global literary market, and critical irrealism, which stands apart in its criticality, asserting a "conscious case of protest both against the bourgeoisie/colonial domination and the hegemonic nature of social realism."[46]

As a critically irrealist fiction that critiques the real extractivist systems ensuring an incessant supply of exploitable labor to the petro-regimes, Unnikrishnan's tale re-creates the popular fantasy of an exploitable humanoid force made ready for capitalistic consumption as seen in several dystopian texts, such as Philip K. Dick's *Do Androids Dream of Electric Sheep?* (later retitled *Blade Runner: Do Androids Dream of Electric Sheep?*), Kazuo Ishiguro's *Never Let Me Go*, and John Brunner's *Into the Slave Nebula*, and in movies like *The Matrix*, *The Surrogates*, Alex Rivera's *Sleep Dealer*, and *Automata*. But what sets Unnikrishnan's MALLUS apart from the conventional tales of robotic human look-alikes is their sheer "naturalness." Unlike the artificially engineered cybernetic humanoids or the AI-infused synthetic androids of the cyberpunk genre, MALLUS are utterly organic, growing as bioengineered hybrid species of fruits and thus as "natural" produce to be harvested: Dr. Moosa's special seeds grew into "oak-dark heat-resistant five-foot-seven Malayalees in twenty three days,"[47] fed by organic fertilizers and grown in soil inside greenhouses. Consequently, the trope of the "natural" in this absurd tale of bioengineered laborers not only offers a critique of capitalistic labor exploitation but signifies an inextricable parallel between human and ecological exploitation under capitalism, "one that turned the work of the human and the nonhuman alike into cheap things."[48] The dehumanization of the humanoid laborer and the simultaneous emphasis on their "organic" state of evolution thus invoke what Patel and Moore read as "capitalism's most sinister accounting tricks,"[49] whereby most humans were put in the category of nature rather than society, facilitating cheap work.

John Bellamy Foster and Brett Clark's reading of the Marxian ideas of ecology are particularly significant in the context of this tale.[50] Foster's notion of the "metabolic rift" is built on Marx's castigation of "capitalist production as a process that disturbs the metabolic interaction between man and the earth."[51] Although Foster's idea was primarily conceptualized in the context of depleting soil fertility, it foregrounds a common coordinate of human and natural exploitation, as Marx asserts:

> All progress in capitalist agriculture is a progress in the art, not only of robbing the worker, but of robbing the soil; all progress in increasing the fertility of the soil for a given time is a progress towards ruining the more long-lasting sources of that fertility. . . . Capitalist production, therefore, develops technology . . . only by sapping the original sources of all wealth—the soil and the worker.[52]

Similarly, Unnikrishnan's irrealist tale blurs the boundaries between human and nonhuman resources, proposing human bodies as part of a system of endless natural extraction. It is here that the agrometaphor assumes particular significance, because the trope of ceaseless harvesting without any exchange price suggests a shift beyond *exploitation*, which Foster and Clark explain as the Marxian notion of "an inner force that propels it" (capitalism) and is governed by the process of abuse of labor power "under the guise of equal exchange," to a disturbing dystopia of *expropriation*, which Foster and Clark define as capitalism's primary relation to its external environment based on the notion of "appropriation . . . without exchange" or "without equivalent."[53] Likewise, Nancy Fraser makes a distinction between the "exploited" wage labor, "the free but propertyless producers who must sell their labor power piecemeal in order to live," and the "expropriated," the "unfree, dependent and unwaged labor."[54] Calling it distinct from "Marxian exploitation, but equally integral to capitalist development," Fraser conceptualizes expropriation as "confiscation-cum-conscription-into accumulation."[55] The driving notion behind expropriation is the assumed lack of agency of nature and certain populations, who along with nature are considered merely "free gifts . . . to capital" to be used and "abused" at will.[56] Patel and Moore assert a similar appropriation as a mode of cheapening labor: "The appropriation—really, a kind of ongoing theft—of unpaid work of women, nature and colonies—is the fundamental condition of the exploitation of labor power in the commodity system."[57]

It is important to recall the concepts of expropriation to understand the significance of Unnikrishnan's irrealist imagery. "In Mussafah Grew People" is a fantasy of a perfect expropriation in which free labor supplies are guaranteed through the easily harvestable humanoid fruits, MALLUS, who just need to be picked, processed, and set to work without wages. The emphasis on the sheer corporeality of the MALLUS labor—bioengineered in the form of "oak-dark, heat resistant five foot seven"[58] perfect bodies—implies a strategic parallel between capitalistic energy regimes and labor exertion. Bellamy and Diamanti recall the Marxian correlative between energy forms and the commodification of human labor power:

> Human exertion becomes a flow of energy in the concrete, while at the same time being modulated by the value form of capital in the abstract. The calorie burners of a human body offer a relatively inefficient source of physical energy compared to even the heat and light released from burning a piece of coal. Yet no lump of coal ever got up and threw itself into the furnace of the steam engine. Capital thrusts human and fossil energy together to extract surplus value from the former but at a greater and greater magnitude due to the energic efficiency of the latter.[59]

The expropriation of the humanoid bodies—reduced to their quintessential muscle power by being "cerebrally customized"—thus suggests a metaphoric correlation between unrelenting labor power and the seeming inexhaustibility of subterranean fossil fuel power, for neither "can produce surplus value independent of the other because each form of energy congeals unevenly into, and is in turn socially regulated by, what Marx calls the 'organic composition of capital.'"[60] The total appropriation of the MALLUS bodies also hinges on their powerlessness in their predicament as disposable bodies of the oil economy's "necropower," or what Mbembe calls "the capacity to define who matters and who does not, who is disposable and who is not."[61] The MALLUS, as the story describes, are "designed to have an average life span of twelve years, after which each would report back to the headquarters like a dying pachyderm . . . , to be driven to the desert for the final chapter in its cycle."[62] The bodies are thus designed for maximum consumption only for a stipulated time, meant to be used up or wasted by what Michelle Yates calls "the logic of human disposability, . . . at accelerated rates in order to secure the most profit."[63] The language of violence is intentionally distant and scientific, normalizing the violence through the rationale of the state's predestined "right to kill"[64] the spent MALLUS bodies. Consequently, when the plan goes awry due to Dr. Moosa's revolt, the oil economy's necropower is unleashed on hundreds of crates of MALLUS seeds that are disposed of in the desert as a method of "management of the multitudes"[65] that have lost their entitlement to life. Rendered as expendable waste when their surplus value is compromised, the MALLUS embody the ultimate capitalistic fantasy of an inexhaustible, disposable, and wage-free labor.

Unnikrishnan's irrealist imagery of the perfectly extractible humanoid body assumes an ominous stature when it is contrasted with the deplorable state of the real Gulf Malayalee petro-precariat. "In Mussafah" makes an apparent distinction between the real-life Malayalees, the wage laborers who work under the semblance of an exchange value, and the MALLUS embodying the expropriated labor. Yet the absurdist tale of human harvest stands as an implied expropriation of the real Malayalee labor in the Gulf, their utter dehumanization pointing at the ground realities of the seemingly voluntary, contractual labor that actually replicates a system of modern slavery. The metaphor coalesces the nonhuman and the human in indistinguishable forms of extraction—as Foster and Clark cite in Marx's reading of slavery, in which "the worker is distinguishable only as *instrumentum vocale* [speaking instrument] from an animal, which is *instrumentum semi-vocale* [semi-mute instrument], and from a lifeless implement, which is *instrumentum mutum* [mute implement]"—suggesting a total erasure of the human agency of slave labor.[66] In Unnikrishnan's text, too, the MALLUS's speech does not testify to human agency but rather works as a camouflage to allow the "unnatural" produce to blend in seamlessly in a petro-economy that thrives on an incessant supply of dehumanized labor. Endowed with only speech and no other identifiable marker of independent human existence, the Malayalees and MALLUS are thus barely distinguishable, conjoined in a common predicament of expropriation.

Finally, as a critique of an existing petro-system, Unnikrishnan's absurdist tale also highlights the racial dimension of the expropriated petro-labor. Central to both Foster's and Fraser's histories of expropriation is also the intricate relationship between capitalistic development and colonial expansion, the enslavement of the populations on the periphery. Fraser asserts the essentially imperialist characteristic of expropriation:

> Far from being sporadic, moreover, expropriation has always been part and parcel of capitalism's history, as has the racial oppression with which it is linked. No one doubts that racially organized slavery, colonial plunder, and land enclosures generated much of the initial capital that kick-started the system's development.[67]

Similarly, Patel and Moore also assert the intricate relationship between cheap labor and "the racial orders by which bodies were read, categorized, and policed at the boundaries of Society and Nature."[68] The overt reference in "In Mussafah" to Keralite labor in the Gulf thus contextualizes petro-capitalism in a postcolonial world. The story exposes Gulf oil capitalism as intricately entwined with the peripheries of colonialism, embodying what I would call a "mature capitalism" that "relies on regular infusions of commandeered capacities and resources, especially from racialized subjects."[69] The petro-precariat is thus reinscribed on a trajectory of India's colonial transnational labor migration—the unfree servitude of indentured laborers that formed a postslavery but nevertheless racialized mode of expropriation. Unnikrishnan's tale thus not only makes visible the violence of a petro-economy and its discreet systems of extraction, enriching the existing "aesthetics of petroleum,"[70] but also facilitates a distinctly postcolonial-materialist petro-critical paradigm potent enough to address the complexity of our contemporary petrosphere.

Notes

1. Patrick Chamoiseau and Ernst van Altena, *Texaco* (Baarn, NL: Ambo, 1994).
2. Karl Marx, *Capital*, vol. 1 (London: Penguin, 1976), 139–54.
3. Deepak Unnikrishnan, "In Mussafah Grew People," in *Temporary People* (Brooklyn, NY: Restless Books, 2017).
4. Michael Lowy, "The Current of Critical Irrealism: A Moonlit Enchanted Night,"

in *Adventures in Realism*, ed. Matthew Beaumont (Oxford, UK: Blackwell, 2007), 193–206.

5. Andreas Malm, *Fossil Capital: The Rise of Steam Power and the Roots of Global Warming* (New York: Verso, 2016), 289.

6. Raj Patel and Jason W. Moore, *A History of the World in Seven Cheap Things: A Guide to Capitalism, Nature, and the Future of the Planet* (Oakland: University of California Press, 2017), 22.

7. Graeme Macdonald, "Containing Oil: The Pipeline in Petroculture," in *Petrocultures: Oil, Politics, Culture*, ed. Sheena Wilson, Adam Carlson, and Imre Szeman (Oakland: University of California Press, 2017), 37.

8. Imre Szeman, "Literature and Energy Futures," *PMLA* 126, no. 2 (2011): 324.

9. Macdonald, "Containing Oil," 36.

10. Amitav Ghosh, *Incendiary Circumstances: A Chronicle of the Turmoil of Our Times* (Boston: Houghton Mifflin, 2005), 138–51.

11. Graeme Macdonald, "Oil and World Literature," *American Book Review* 33, no. 3 (2012): 7, 31.

12. Rosalind Williams, *What Is a Petroculture? Conjectures on Energy and Global Culture*, October 11, 2017, video, MIT, https://www.youtube.com/watch?v=CcPq6W9Psgs.

13. Macdonald, "Oil and World Literature," 31.

14. Ibid.

15. Imre Szeman, "Conjectures on World Energy Literature: Or, What Is Petroculture?," *Journal of Postcolonial Writing* 53, no. 3 (2017): 285.

16. Macdonald, "Oil and World Literature," 31.

17. Ibid.

18. Ibid.

19. Brent Ryan Bellamy and Jeff Diamanti, "Phantasmagorias of Energy: Toward a Critical Theory of Energy and Economy," *Mediations* 32, no. 1 (Spring 2018): 12.

20. Victor Turner, *The Forest of Symbols: Aspects of Ndembu Ritual* (Ithaca: Cornell University Press, 1967), 96.

21. Jean-François Mouhot, "Thomas Jefferson and I," *Solutions* 5, no. 6 (November 2014): 75.

22. Vani Saraswathi, "Living an Unreal Dream," *The Hindu*, April 15, 2017, https://www.thehindu.com/books/books-reviews/living-an-unreal-dream/article18058454.ece.

23. Goran Bregović, "Gas, Gas," Spotify, track 2 on *Goran Bregović's Karmen with a Happy End*, 2016, https://open.spotify.com/album/1msJUuUc3SkddfMFogoUXb?highlight=spotify:track:50Kos51iL8QzaL72EBuYKq.

24. Ghosh, "Petrofiction," 142.

25. Sheena Wilson, Adam Carlson, and Imre Szeman, eds., *Petrocultures: Oil, Politics, Culture* (Oakland: University of California Press, 2017), 37.

26. Touraj Atabaki, Elisabetta Bini, and Kaveh Ehsani, *Working for Oil: Comparative Social Histories of Labor in the Global Oil Industry* (London: Palgrave Macmillan, 2018), 23.

27. Ibid., 4.

28. Michael Rubenstein, "Petrocriticism," in *Futures of Comparative Literature: ACLA State of the Discipline Report*, ed. Ursula Heise (New York: Routledge, 2017), 50.

29. Andreas Malm, "The Origins of Fossil Capital: From Water to Steam in the British Cotton Industry," *Historical Materialism* 21, no. 1 (2013): 33.

30. Carl Skutsch, ed., *Encyclopedia of the World's Minorities* (New York: Routledge, 2013), 586.

31. The London School of Economics published a report on the system for governing and regulating migration. Although the *kafala* has roots in both law and custom, in practice it consists of three basic features: "It establishes that entry for the purposes of work requires a local sponsor; it establishes the sponsor's responsibility for the sponsored migrant's housing, employment conditions, and other benefits; and it establishes that the migrant's exit and the migrant's capacity to change employers is subject to the sponsor's permission."

32. I borrow Guy Standing's term "precariat," which signifies a common

predicament of disposability and labor itinerancy. Standing, *The Precariat: The New Dangerous Class* (London: Bloomsbury, 2019).

33. W. Kurt Hauser, *Invisible Slaves: The Victims and Perpetrators of Modern-Day Slavery* (Stanford, CA: Hoover Institution Press, 2018), 31.

34. Andrew Nikiforuk, *The Energy of Slaves: Oil and the New Servitude* (Vancouver: Greystone Books, 2014).

35. Mouhot, "Thomas Jefferson and I," 75.

36. Ibid.

37. Unnikrishnan, "In Mussafah Grew People," 37.

38. Ibid.

39. Speaking of the popular image of the *Gulf-karan*, S. Sanandakumar explains that "the old Gulf Malayalee was a neo rich, semi-literate upstart who lands up at the airport sporting dark glasses and bell-bottom trousers with a National Panasonic NSE, two-in-one tape recorder in one hand and in the other a suitcase full of perfume, liquor and other goodies for his friends and relatives." "A Fifty Year Old Phenomenon Explained: Malayalee Migration to Gulf Builds the New Kerala," *Economic Times*, October 13, 2015, https://economictimes.indiatimes.com/news/politics-and-nation/a-fifty-year-old-phenomenon-explained-malayalee-migration-to-gulf-builds-the-new-kerala/articleshow/49201357.cms?from=mdr.

40. Mohamed Shafeeq K., "The Gulf on the Malayali Big Screen: An Outline History," *Refugee Watch Online*, July 16, 2016.

41. Lowy, "Current of Critical Irrealism," 193.

42. Ibid., 205.

43. Ibid., 196.

44. Ibid.

45. Ibid.

46. Sourit Bhattacharya, "The Margins of Postcolonial Urbanity: Reading Critical Irrealism in Nabarun Bhattacharya's Fiction," in *Postcolonial Urban Outcasts: City Margins in South Asian Literature*, ed. Madhurima Chakraborty and Umme Al-wazedi (New York: Routledge, 2017), 41–42.

47. Unnikrishnan, "In Mussafah Grew People," 37.

48. Patel and Moore, *History of the World*, 61.

49. Ibid., 93.

50. John Bellamy Foster and Brett Clark, "The Expropriation of Nature," *Monthly Review* 69, no. 10 (2018).

51. Marx, *Capital*, 637–38.

52. Ibid.

53. Foster and Clark, "Expropriation of Nature."

54. Nancy Fraser, "Expropriation and Exploitation in Racialized Capitalism: A Reply to Michael Dawson," *Critical Historical Studies* 3, no. 1 (Spring 2016): 166–67.

55. Fraser mentions a variety of forms of expropriation: "The confiscation may be blatant and violent, as in New World slavery—or it may be veiled by a cloak of commerce, as in the predatory loans and debt foreclosures of the present era. The expropriated subjects may be rural or indigenous communities in the capitalist periphery—or they may be members of subject or subordinated groups in the capitalist core. They may end up as exploited proletarians, if they're lucky—or, if not, as paupers, slum dwellers, sharecroppers, 'natives,' or slaves, subjects of ongoing expropriation outside the wage nexus. The confiscated assets may be labor, land, animals, tools, mineral or energy deposits—but also human beings, their sexual and reproductive capacities, their children and bodily organs. The conscription of these assets into capital's circuits may be direct, involving immediate conversion into value—as, again, in slavery; or it may be mediated and indirect, as in the unwaged labor of family members in semi-proletarianized households."

56. Patel and Moore, *History of the World*, 93.

57. Ibid.

58. Unnikrishnan, "In Mussafah Grew People," 37.

59. Bellamy and Diamanti, "Phantasmagorias of Energy," 4.

60. Ibid.

61. Achille Mbembe, "Necropolitics," *Public Culture* 15, no. 1 (Winter 2003): 27.

62. Unnikrishnan, "In Mussafah Grew People," 38.

63. Michelle Yates, "The Human-as-Waste, the Labor Theory of Value and Disposability in Contemporary Capitalism," *Antipode* 43, no. 5 (2011): 11681. Yates asserts that "intrinsic to capitalism's generation of objective waste is the logical necessity of wasting human lives. For labor is only one of the objective factors of production and is, like any of its factors, expendable if its expenditure bolsters a rate of profit. These two modalities of by-product—the redundant worker and the residuals from the production process itself—demonstrate the necessary relation between the generation of (surplus) value and the generation of surplus populations (as a form of waste)."

64. Mbembe, "Necropolitics," 16.

65. Ibid., 34.

66. Foster and Clark, "The Expropriation of Nature."

67. Fraser, "Expropriation and Exploitation in Racialized Capitalism," 167.

68. Patel and Moore, *History of the World*, 196.

69. Ibid., 167.

70. Stephanie LeMenager, "The Aesthetics of Petroleum, After Oil!" *American Literary History* 24, no. 1 (Spring 2012): 59–86.

9.

Petro-cosmopolitics

Oil and the Indian Ocean in Amitav Ghosh's *The Circle of Reason*

Micheal Angelo Rumore

A memorable scene from Amitav Ghosh's debut novel, *The Circle of Reason* (1986), depicts the journey of a diverse party of migrants from Kerala to the fictional Gulf state al-Ghazira aboard a ramshackle ship called the *Mariamma*. As its newly installed diesel engine sputters in the middle of the Indian Ocean, the ship's owner, Hajji Musa, pauses to reflect on "the stories people told up and down the Malabar Coast of boats setting off for al-Ghazira with twenty, forty and even (so they said) a hundred eager emigrants, but only to run out of fuel halfway, or else be swallowed in the sea with the first mild gale, borne down by sheer weight."[1] Luckily, the *Mariamma*'s troupe of oil pilgrims arrives safely (at least for the moment) at the Gulf, setting the scene for Ghosh's first foray into what he would six years later, in a famous review of two novels by Abdelrahman Munif, call "petrofiction."[2] The *Mariamma*'s oceanic crossing introduces another trope that also weaves through Ghosh's work: that of the cosmopolitan Indian Ocean imaginary notably developed in his hybrid travelogue, *In an Antique Land: History in the Guise of a Traveler's Tale* (1992), published the same year as the aforementioned review, "Petrofiction: The Oil Encounter and the Novel." Both of these works have had an outsized influence, almost singlehandedly catalyzing two literary fields: petroculture studies and Indian Ocean studies.

But the inception of these two influential strands of Ghosh's work with the background noise of a stammering, marginally functional diesel engine also elicits a sobering realism. The collection of passengers aboard the *Mariamma* may be irreducibly cosmopolitan, but their movement depends on the whims of a temperamental diesel engine—a figure of the broader imperial regimes undergirding the global oil economy. To push Ghosh's metaphor further, oil powers the engine, yes, but all to carry the "lucky" migrants toward the illusory opulence of the oiltown, built—as we will see—on unsteady sand.

In this essay, I read the conjoined articulation of petrofiction and Indian Ocean aesthetics in Ghosh's first and most underexplored novel against the romanticized variant of cosmopolitanism his work is often characterized as positing. Rather, Ghosh's depiction of the simultaneous cosmopolitanism and exploitation of Indian Ocean labor migration to the oil-producing Persian Gulf states anticipates his argument in "Petrofiction" that oil poses grave representational challenges for both American and postcolonial literatures. I also read the novel in relation to the relative paucity of considerations of oil in Indian Ocean studies today. Fossil fuels, I contend, have had an underaccounted-for role in producing the Indian Ocean as a cosmopolitical site. Thus, I position *The Circle of Reason*'s original entangled depictions of oil and the Indian Ocean as a challenge to the relative separation of cosmopolitics from the imperial dynamics of the global oil economy. Since more than half of the world's oil crosses the Indian Ocean, this split exposes how oil continues to slip through the cracks of our theoretical and historical languages.

If, as Ghosh memorably put it in "Petrofiction," oil "smells bad," its rancid odor also seems to repel the cosmopolitical ambitions of Indian Ocean studies.[3] This repulsion stems, I would argue, from a number of reasons related to the field's investment in tracing "premodern" cosmopolitan connections as distinct from the "modern," anticosmopolitan (because nationalist) present. Take, for example, *In an Antique Land*, Ghosh's hybrid travelogue that attempts to retrace the oceanic travels of a medieval Jewish merchant's Indian slave, a difficult endeavor in the present given the partitioning of such fluid Indian Ocean worlds by the long histories of imperialism and nationalism. Isabel Hofmeyr credits *In an Antique Land* in particular with introducing a "comparative" logic to Indian Ocean studies, succinctly summarized as "cosmopolitanism then, nationalism now."[4] By extension, the oil encounter must be located in the nationalist present, superimposed on the cosmopolitan past. Politically, the notion of the Indian Ocean as a cosmopolitan site both before and outside of Western colonialism acts as a reversal of the teleology of globalization: from the vantage point of the Indian Ocean, Western colonial intervention resulted not in the positing of a more "connected" world but in the demonstrable enclosure of one. Again, Ghosh's *In an Antique Land* is an exemplary text in this regard: one recalls its description of Portuguese warships imposing the "gifts" of colonial modernity—epistemologies of nation, race, and the state—on the multilateral sovereignties that characterized the Indian Ocean prior to their arrival. Thus, narrating the enclosure of preexisting, non-Western forms of globalization that link the Middle East, South Asia, and East Africa also becomes a strategy of rendering and ultimately challenging the epistemic order—and material violence—of modern coloniality.

Of course, cosmopolitanism itself remains a highly contested subject, and doing justice to the debates on its provenance will require a scope far beyond that of this essay. Numerous postcolonial critiques and variations of the term abound, generally resulting in a rethinking of cosmopolitanism as not an abstract universalism but as particular, locatable political practices, which Pheng Cheah and Bruce Robbins have called "cosmopolitics."[5] In Indian Ocean studies, however, the pertinent question has been the accuracy of the comparative logic that has resulted from designating the premodern Indian Ocean as cosmopolitan—a logic, as previously mentioned, often traced back to Ghosh's literary work. As Sugata Bose points out, one negative consequence of this stark division between "premodern" and "modern" has been the assumption that the Indian Ocean "stopped being a system or arena around the mid- to late eighteenth century," a notion we might link to the absence of oil in such work.[6] Other Africanist scholars, such as Kai Kresse and Edward Simpson, have critiqued the "lazy" way the adjective "cosmopolitan" tends to be affixed to the Indian Ocean, a practice that they argue evacuates the region of the "historical struggles on which it builds."[7] Motivated by similar worries, Gaurav Desai relates this debate directly back to Ghosh's *In an Antique Land*, arguing that the text represents the "production of history in a nostalgic mode," one that romanticizes an ahistorically "non-coercive" past in opposition to modern violence.[8] While I largely share these concerns about the dangers and limits of ahistorically employing "cosmopolitan" as an adjective, I also want to argue that what I am calling Ghosh's cosmopolitics can be read as a critique of the kind of nostalgia that Desai rightly identifies. I have come to this conclusion in part by putting Ghosh's petrofictional critique in conversation with his Indian Ocean aesthetics.

I maintain that Ghosh's cosmopolitics are neither simply nostalgic nor incompatible with historical realism. Rather, I read the appeals to oceanic cosmopolitanism throughout Ghosh's work as grappling with the notion that the irreducibly transcultural lifeworlds of the premodern Indian Ocean cannot be translated into the modern dictates of identity. Ghosh's work is therefore cosmopolitical in the sense that it is committed to transgressing the epistemic order of modernity as a crucial postcolonial project—one Ghosh relates directly to questions of novelistic form. In any case, my use of these terms follows from this ethical stance. Clarifying this cosmopolitical orientation is important, I contend, because the mismatch between the diverse human experiences of globalization and the narrative convention that Ghosh identifies in his Indian Ocean work is essentially the same representational problem he confronts in relation to the oil encounter. In the case of the oil encounter, particularly as represented in *The Circle of Reason,* the opposition between cosmopolitanism and modern violence is not so clearly a relationship of past and present but a clash of temporal affiliations in the present. What emerges is closer to Ngũgĩ wa Thiong'o's dialectical reading of cosmopolitanism, in which the (post)colonial figures as a "depository of the cosmopolitan" defined as "bearing the marks of many streams," in contrast to the anticosmopolitan violence of the colonial *Bildung.*[9] Oil figures here as a lubricant of modernity that simultaneously enables cosmopolitical resistance and reinscribes these possibilities within a colonial epistemic order that renders them illegible. Think back to the diesel-powered ships in *The Circle of Reason*—making movement possible but sputtering along in the Indian Ocean, carrying migrants toward an existence as disposable labor in the false New World of the Gulf oil economy. The novel, as I will show, treats this experience with little nostalgia.

In retrospect, however, the characterization of Ghosh's Indian Ocean work as "nostalgic" strikes this reader as ironic. After all, Ghosh himself notably uses the word pejoratively in "Petrofiction." Toward the essay's conclusion, Ghosh describes the climax of *Cities of Salt*—in which oil workers successfully stage a revolution against American imperialism and its neocolonial avatars—as an "escapist fantasy" relying on "a kind of nostalgia, a romantic hearkening to a pristine, unspoiled past."[10] Critics of Ghosh's Indian Ocean aesthetics have used strikingly similar words. How is it, then, that Ghosh imagined an Indian Ocean aesthetics now itself called nostalgic by both proponents and critics as an antidote to such an "escapist fantasy"? Whereas Peter Hitchcock takes exception to the apparent banishment of working-class victory to the realm of the fantastic, I read Ghosh as concerned less with representation in the positive sense and more with the epistemic conditions by which the oil encounter (and, by extension, resistance to it) can be narrated—specifically in the form of the novel.[11] Thus, at least on one level, Ghosh's more recent pessimism in *The Great Derangement* (2016) over the novel as a vehicle for representing the climate crisis recasts an aesthetic problem present throughout Ghosh's literary career and implicit in his early attempts to engage with the oil encounter in *The Circle of Reason* and subsequent works—namely, that of the noncoincidence of the subaltern experiences produced by the contemporary petrosphere and the nationalist orientation of novelistic conventions. Put slightly differently, the "derangement" Ghosh speaks of refers not only to narrative absences but also to failures of narration. This cosmopolitics will become clear from close readings of both "Petrofiction" and *The Circle of Reason*. But as I will argue, the comparative logic of Ghosh's cosmopolitics articulates a humanist aesthetics that is substantively different from the postcolonial grammar he is often associated with.

"Petrofiction" has often been cited as a critique of the insularity of American fiction and an open call for the "Great American Oil Novel." In Hitchcock's words, the essay "threw down a gauntlet, a challenge to find and elaborate a significant counter-critique of petrofiction rooted in the specific geo-political circumstances of the US's rise to global power over the last century."[12] But while "Petrofiction" did indeed generate a series of "petrocultural" investigations and critiques within American studies (a fact Ghosh himself responded to with "very great surprise"), the essay was not only concerned with American fiction. Oil may "smell bad" to American literary noses, but Ghosh continues the essay by arguing that the literary imagination on the "Arab side" considers oil equally malodorous. This, Ghosh claims, is because of a general marginalization of "messy" littoral histories within "high" Arabic literary traditions:

> Until quite recently, the littoral of the Gulf was considered an outlying region within the Arab world, a kind of frontier whose inhabitants' worth lay more in their virtuous simplicity than in their cultural aspirations. The slight curl of the lip that inevitably accompanies an attitude of that kind has become, if anything, a good deal more pronounced now that many Arab writers from Egypt and Lebanon—countries with faltering economies but rich literary traditions—are constrained to earn their livelihood in the Gulf. As a result, young Arab writers are no more likely to write about the Oil Encounter than are their Western counterparts.[13]

Thus, Ghosh presents the American and Arab sides as two sides of the same coin of postcolonial modernity and its epistemic orderings. The oil encounter, consequently, is absent from the literary imagination as a result not only of

American myopia but also of the diasporic circuits that the oil economy gives rise to. Crucially, Ghosh emphasizes the "physical and demographic separation of oil installations and their works from the indigenous population."[14] Oil imperialism may have produced "city-states where virtually everyone is a 'foreigner'; an admixture of peoples and cultures on a scale never before envisioned," but it has also produced "vicious systems of helotry juxtaposed with unparalleled wealth; deserts transformed by technology and military devastation on an apocalyptic scale."[15] Therefore, even if we attempt, as Michael Walonen usefully does, to articulate a petrofictional world-literary countercanon, we would sidestep Ghosh's primary question, which is an aesthetic one: how to represent the simultaneous cosmopolitanism and rigidly enforced taxonomies of difference posed by the oil encounter.[16] In other words, Ghosh is troubled by the entrapment of the oil encounter within what Denise Ferreira da Silva would call the "modern scene of representation"—the epistemic order by which the colonial *Bildung* becomes instituted as the signifier of globality—but at the same time he seeks to work within the novel form to depict how the everyday lives of oil migrants constantly exceed and resist this order.[17]

The problem of representing the oil encounter thus becomes less *whether* it appears but under what epistemological, and therefore narrative, conditions it can appear. Ghosh explains:

> The experiences that oil has generated run counter to many of the historical imperatives that have shaped writing over the last couple of centuries and given it its distinctive forms . . . , such notions as the idea of distinguishable and distinct civilizations, or recognizable and separate 'societies.' It is a world whose closest analogues are medieval, not modern—which is probably why it has proved so successful in eluding the gaze of contemporary global culture. The truth is that we do not yet possess the form that can give the Oil Encounter a literary expression.[18]

In part, Ghosh points out the representational limits of methodological nationalism, the assumption that nations follow discrete, self-contained historical trajectories. In the realm of literary production, of course, the novel has long been connected to the modern nation as, in Benedict Anderson's famous phrase, an "imagined community." As I have written in another context, the European novel emerged in relation to tensions over "translatability" between a classical humanist universalism and a burgeoning notion of literary production as the expression of national cultural essences.[19] Ghosh, in his setting, is concerned with a different question of translatability: the postcolonial problem of rendering transcultural histories within the conventions of novelistic form, which not only maintains notions of discrete national and "civilizational" literary traditions but also has an inescapably bourgeois orientation toward linear progress that shares a constitutive logic with the kinds of erasures identified in "Petrofiction." Ghosh thus implicitly links novelistic form to neocolonial notions of national "development," a connection explored throughout his work in relation to commodities other than oil—for example, textiles in *The Circle of Reason*, teak and rubber in *The Glass Palace* (2000), and, of course, opium in the *Ibis Trilogy* (2008–15).

By calling attention to these stratified but still cosmopolitical diasporic networks, Ghosh also gestures toward epistemic concerns. Ghosh's remark that the oil encounter's closest analogies are "medieval, not modern" signals this stance: the point is less that oil is "not modern" than that modernist narrative assumptions—particularly in the relation of the novel's bourgeois form to national development claims—render the irreducibly cosmopolitan conditions of the oil economy (what Edward Said once called the "brute reality" existing beneath orientalist lenses) illegible to modern eyes and, hence, modern pens. The "medieval" comparison, of course, also alludes to the premodern Indian Ocean worlds that would provide context for many of Ghosh's works—most notably *In an Antique Land* but including *The Circle of Reason*, to which I will now turn. Ghosh's debut novel, therefore, will be read as preempting the critique of what he takes to be Munif's "nostalgia" for "a pristine, unspoiled past," a critique I read as a concern over how the novel form reproduces the representational order of colonial modernity.

Following the "potato-headed" orphan Alu from Bengal to the Persian Gulf to North Africa, *The Circle of Reason* explores the complex entanglements of cosmopolitan histories and neocolonial appropriations. The first section of the novel, "Satwa: Reason," depicts Alu's relationship with his foster father, Balaram, whose obsession with the *Life of Louis Pasteur* leads him to fetishize notions of purification, ultimately resulting in a scheme to douse the village in large quantities of carbolic acid. Thus, the text introduces its eponymous "circle of reason." As other interpreters have noted, the link between reason and purification belies a debilitating Eurocentrism. Tuomas Huttunen, for example, is right to note how "reason is linked in the narrative with the idea of the purity of the poles in Western binary constructions."[20] But this imperialistic conception of rationality is not the only sense in which reason is discussed in the text. The novel as a whole is implicitly compassionate toward the spirit of Balaram's "circular" reason, even as it lampoons the institutionalization of his ideas in his Rationalist club and School of Reason. And yet reason's "circle" also represents something of a double bind: reason is depicted not as just a linear encroachment of colonial rationality but more ambivalently. For example, Balaram's intellectual antagonist, Bhudeb Roy, insists on reason as a "straight line," opposing Balaram's circular reason with a kind of developmentalist logic. "Look at Europe, look at America, look at Tokyo," Roy says. "Straight lines, that's the secret."[21] Balaram opposes this teleological conception of reason with a more cosmopolitan one. An obvious example can be seen in Balaram's attempts to reconcile "Hindu sages and modern science," denying the West's exclusive claim to reason at the same time he humorously declares a program to eliminate dirty underpants.[22] For all of Balaram's naïveté, the novel remains sympathetic to his deterritorialized notion of reason. At the same time, Ghosh is clearly aware of the conscription of ideologies of rationality by (neo)colonial political forms and their aggressively territorializing trades in commodities and labor power.

This "bloody irony," as the text puts it, first appears in relation to the circulation of textiles, as we learn of Alu's precocious mastery of the art of weaving. In the world of the text, weaving also represents an idealized mastery of reason. The loom, we are told, "has never permitted the division of reason."[23] The following genealogy of the ancient proliferation of Indian cloths is perhaps the first description of cosmopolitan Indian Ocean trade appearing in Ghosh's work: "All through those centuries cloth, in its richness and variety, bound the Mediterranean to Asia, India to Africa, the Arab world to Europe, in equal, bountiful trade."[24] Still, even though the novel vaunts a deterritorialized version of "mechanical man," personified in the organic practice of weaving, it also narrates the tragic appropriation of the loom's autochthonous cosmopolitanism by colonial rationalization, described as the end of a world: "When the history of the world broke, cotton and cloth were behind it; mechanical man in pursuit of his own destruction."[25] Here the cosmopolitan history of cloth gives way to the violent coloniality of the cotton trade. As a whole, the novel refuses to resolve these two contradictory, superimposed conceptions of reason—in fact, it reads them dialectically.

I would argue that critics who have found the novel incoherent have failed to take this insistent ambivalence seriously because of theoretical assumptions that reason should be identified solely with colonial teleologies. Rather, Ghosh holds the double bind of reason in suspension:

> Every scrap of cloth is stained by a bloody past. But it is the only history we have and history is hope as well as despair. . . . And so weaving, too, is hope; a living belief that having once made the earth one and blessed it with its diversity it must do so again. Weaving is hope because it has no country, no continent. . . . Weaving *is* Reason, which makes the world mad and makes it human.[26]

Balaram's arc is similarly "mad" and "human." Despite his deterritorialization of reason, his cosmopolitanism exists in tension with an obsession with purification, which leads to a standoff with the authorities, who take his School of Reason and stockpiling of carbolic acid as an extremist threat. The violent climax of Balaram's pasteurizing schemes throws Alu into motion, as he flees to al-Ghazira to avoid capture by the bird-watching policeman Jyoti Das. As a result, Alu retraces the cosmopolitan routes attributed to the cloth trade, only now these routes support another commodity: oil. Thus, the novel becomes an oil fiction in earnest.

The descriptions of the al-Ghazira oiltown in the second part of the novel, "Rajas: Passion," continue to develop the themes of reason and purification introduced in the first section. After finding passage from Kerala to al-Ghazira under the auspices of the Egyptian madam Zindi, Alu works in construction until the collapse of the towering Star skyscraper—a symbol of al-Ghazira's gilded prosperity—buries him beneath a pile of rubble, which he miraculously survives. Stephanie Jones usefully notes the "polylingualism" of this part of the novel, reading in Deleuzian terms the resistant minor languages that abound in the depiction of al-Ghazira.[27] The gossip swirling around Alu's apparent survival provides the backdrop for recounting the oiltown's formation, a moment in the novel also indicative of the tragic reinscription of these cosmopolitan voices within a kind of neocolonial epistemic ordering. Alu's survival appears initially in the form of rumors of his voice having been heard crying out from the rubble. At this point, Ghosh focuses less on the objective perspective of the narrator and more on the dialogue between al-Ghazira's many minor characters. Through this multitude of fleeting voices, we receive contending accounts both of the Star's collapse and of the history of al-Ghazira as a petro-state. In one of the few essays to give sustained attention to *The Circle of Reason* specifically as an oil fiction, Claire Chambers notes the formal mélange that emerges here, which mixes elements of the picaresque, detective fiction, social realism, and magical realism.[28] These multigeneric qualities are displayed in competing explanations of the Star's collapse from two minor characters, Abu Fahl and Hajj Fahmy.

Abu Fahl provides the pithy social realist account, claiming that there is "no mystery" to the Star's sudden disintegration: "Everyone knows that the contractors and architects put too much sand in the cement. . . . A cement shortage, they say. But actually they're busy putting up palaces with the money they make from that cement—for themselves at home in England, or India, or Egypt, America, Korea, Pakistan, who knows where? The cement they were using for the Star was nothing but sand."[29] The illusory quality of the material underpinnings of the Star ("nothing but sand") clearly also applies to the petro-state itself—a kind of symbolic depiction of al-Ghazira's "universal foreignness" (to adapt the concept Ghosh employs in "Petrofiction"). In addition, the cosmopolitan character of al-Ghazira is linked explicitly to, rather than positioned against, the very different cosmopolitanism of capital: the cement pinched from Ghaziri contracts literally contributes to material foundations located elsewhere. For Abu Fahl, the story is that simple.

In contrast to this cold and cynical realism, Hajj Fahmy spins a more magical realist yarn, which also contains the novel's primary account of the colonial creation of the oiltown. "Many years ago, so long that time is of no significance," Hajj Fahmy relates, there lived an Egyptian egg-seller called Nury, who was "so painfully cross-eyed . . . that whenever other people only saw Cairo he could see Bombay as well."[30] As the primary egg supplier, Nury uniquely overhears the gossip of all sectors of Ghaziri society. In the background, the discovery of oil leads the British to attempt to force an oil treaty on the reigning Malik. Under the threat of warships and Indian soldiers, the Malik relents on the condition that "the Oilmen never leave the Oiltown and never enter al-Ghazira."[31] The egg-seller, however, discovers a plot to circumvent this agreement by deposing the Malik and replacing him with his European-educated (and petro-friendly) brother, the Amir. The scheme centers on the importation of "specially grown date palms" that "could thrive on any soil, however inhospitable," which presumably would stir the people to support British intervention.[32] Thus, we see an intersection of ecological imperialism (the simultaneous promise and threat of "magically" engineered seeds) and what Brian Larkin calls the "colonial sublime" (the provocation of feelings of "wonder" and "awe" by the technologies of colonial rule).[33] This detail links what I've called epistemic enclosure to the material processes of ecological enclosure as a strategy of colonial accumulation.

The story continues: "The Malik was bound to resist . . . , [b]ut by then the townspeople, so long loyal to the Malik, would hesitate, dazzled by their glimpse of the Amir's power to turn the desert green, and in the end would rally to his side."[34] The egg-seller relays what he heard to a local merchant, who responds by dousing the magical dates in oil and striking a match, thus "sending the Amir's dreams up in flames."[35] Ecological imperialism thus gives way to petro-imperialism. In response to the resulting commotion among the townsfolk, the oiltown guards—consisting of "Filipino faces, Indian faces, Egyptian faces, Pakistani faces, even a few Ghaziri faces, a whole world of faces"—deploy tear gas on the confused crowds.[36] The egg-seller, however, is the only casualty: an ill-tempered camel bites off his head as he blindly stumbles through the unrest. This lone death, however, symbolizes the broader enclosure of autochthonous trades and their preexisting networks of exchange. The narrator reminds us that "there was no place in [the new Ghazira] for sharp-eyed egg-sellers."[37] In the aftermath, the Amir comes to power, inaugurating the petro-state in earnest, the "whole country" now transformed into an oiltown.

This "magical" history, Hajj Fahmy concludes, more fully explains the collapse of the Star. Built on the charred remains of the bioengineered dates, the Star fell simply because "no one wanted it": "a house which nobody wants cannot stand."[38] In Hajj Fahmy's rendition, the Star's collapse becomes another symbol of the "universal foreignness" of the petro-state. This allegorical resonance is reflected in Hajj Fahmy's descriptions of the oiltown itself, which chronicle the multiple uprootings involved in its formation.

> In those days many Ghaziris wanted work. But there was no work for them in the Oiltown, for the Oilmen knew that a man working on his own land has at least a crop to fight for. They were welcome: since the beginning of time al-Ghazira has been home to anyone who chooses to call it such—if he comes as a man. But those ghosts behind the fence were not men, they were tools—helpless, picked for their poverty. In those days when al-Ghazira was a real country they were brought here to slip between its men and their work, like whiffs of an opium dream; they were brought as weapons, to divide Ghaziris from themselves and the world of sanity; to turn them into buffoons for the world to laugh at.[39]

A number of things in this passage stand out: for one, the projection of a cosmopolitan past onto al-Ghazira from "the beginning of time" until ruptured by the founding of the oiltown. But, importantly for my argument, the enclosure of this cosmopolitan past comes not only through the superimposition of nationalism understood as a unitary claim on identity but also through diffuse processes of differentiation, dehumanization, and dispossession—productive both of resistant cosmopolitan histories and forms of nostalgia that Ghosh remains ambivalent about. We discover something similar to what Michael Watts points out in the context of oil imperialism in the Niger Delta: "What we have in other words is not nation building—understood in the sense of a governable space—but perhaps its reverse: the 'unimagining' or deconstruction of a particular sense of national community."[40] Or perhaps we encounter a kind of nation-building precisely through processes of "unimagining," that is, through the production of rigidly ordered difference by the epistemic and material enclosure of cosmopolitan possibilities—the dark mirror of postcolonial theoretical celebrations of the ontological creativity of migrancy. Ghosh implicitly recognizes oil's corrosive aspects as an aesthetic problem: as mentioned earlier, he notes in "Petrofiction" how the oil encounter "runs counter to many of the historical imperatives that have shaped" modern literary forms.[41] The novel form—which links individual growth to national development—would therefore also run counter to oil's unimagined communities. Just as the Star collapses, possibilities for a sense of shared community collapse. But this in itself does not necessarily foreclose more radically cosmopolitan solidarities, as Alu's rise from the rubble shows.

Upon his emergence from the ruins of the Star, Alu rather unwittingly ignites a mock socialist revolution, which can be read in conversation with Ghosh's later critique of *Cities of Salt* in "Petrofiction." Taking up Balaram's "call to reason," Alu identifies the ultimate spreader of germs and enemy of purity: money. In an uncharacteristically self-possessed moment, Alu declares the necessity of a "war on money," a speech described as magically cosmopolitan:

> I saw that crowd absolutely silent, listening to a man, hardly more than a boy, talk, and that, too, not in one language but in three, four, God knows how many, a khichri of words; couscous, rice, dal, and onions, all stirred together, stamped and boiled, Arabic with Hindi, Hindi swallowing Bengali, English doing a dance; tongues

> unraveled and woven together—nonsense, you say, tongues unraveled are nothing but nonsense—but there again you have a mystery, for everyone understood him, perfectly, like their mother's lullabies. They understood him, for his voice was only the question; the answers were their own.[42]

Thus, Alu articulates a radically cosmopolitan politics but one that inherits the limitations of Balaram's ironic notion of purification. Despite the culinary metaphors employed in this passage, we do not find an example of Rushdiean "chutnification"—an aesthetics of postcolonial hybridity—but precisely the opposite: the war on money instead represents a protest against "chutnified" conditions, ambivalently combining cosmopolitan solidarities and a dismantling of difference under the guise of eliminating germs.

Almost immediately, Alu again becomes a passive protagonist, as his ideas are given form by the migrant workers themselves, resisted only by Zindi and Jeevanbhai Patel, an Indian shopkeeper whose presence in al-Ghazira predates the oil economy. Importantly, it is actually Patel who narrates this section of the novel as it is relayed to Jyoti Das and his diplomatic contact Jai Lal, both of whom dismiss Patel as unreliable. Das, for his part, finds the magical aspects of Patel's depiction unrealistic, while Lal merely worries about the fact that Ghaziri authorities may read the war on money as an Indian uprising, thus interrupting remittances back to India.[43] Here Ghosh presents us with two interrelated failures of both narration and epistemology: both authorities circumscribe the war on money within understandably nationalist narrative norms. It is also significant that Patel and his shop predate the oil regime, as the war on money in part targets the shopkeepers and interrupts their trade. As the migrant laborers begin dousing the shops in carbolic acid, Patel attempts to sell the trade to Zindi, but before the transaction can be formalized, he is arrested and subsequently commits suicide. Falsely linked to the demonstrations because of his long history in al-Ghazira and his loose association with the former regime, Patel is mistaken by the Amir as the ringleader of a coup attempt by the deposed Malik. The demonstrating migrants are then put down violently, leading to a new round of accumulation by the oil regime. Now trailed by Das and targeted by the regime, Alu and Zindi head west toward North Africa, bringing the al-Ghazira section of the novel to a close. Ultimately, the polyvocality characterizing this section of the text is easily absorbed by the material and epistemic violence of the petro-state.

Because of Ghosh's concern for narration and epistemology throughout the novel, I read his portrayal of the multiculturalism of the oil regime as distinct from cosmopolitanism in the sense that the oil economy depends on rigid management of difference that undergirds the racialized division of labor. Previously, I mentioned Hofmeyr's description of Ghosh's Indian Ocean juxtaposition of hybrid cosmopolitan pasts and the monocultural nationalist present, but in Ghosh's depiction of the al-Ghazira petro-state, the nationalism of the present cannot be described as a unitary claim on identity. It is not that one temporality is multicultural and the other is not but that Ghosh distinguishes between a rigid hierarchical assemblage of difference that characterizes the biopolitics of colonial power (which encompasses but is not limited to nationalism) and a more entangled ecology of difference outside the representational order of modernity (what I have described as cosmopolitanism). As in "Petrofiction," the nationalized present is always already globalized in terms of a colonial-taxonomic ordering of difference—multicultures brought forth by the oil economy that obscure cosmopolitan pasts and possible futures. The formation of the oiltown follows, as Reem Alissa describes in another context, the characteristic "combination of socio-spatial segregation and capitalist motivations [that] was a staple of urban development in other oil camps and towns such as those in Venezuela, Iran, Bahrain, and Saudi Arabia."[44] *The Circle of Reason*—in portraying the contradictions between cosmopolitan resistance and the forms of nostalgia produced by the oil economy—links the limits of novelistic representation to the antinomies of political representation. Part of my point here is that Ghosh's cosmopolitics in fact runs counter to the politics of difference that was then being institutionalized—at least in its more postmodern guises—as postcolonial theory, because Ghosh worries about the reinscription of a politics of difference within the representational order of modernity. These ideas are perhaps more clearly developed in *In an Antique Land*, but the insistently non-Romanticized depiction of al-Ghazira, when put in conversation with Ghosh's work as a whole, chastens the nostalgic characterization of his cosmopolitics.

Reason comes full circle in the third section, "Tamas: Death," when Alu has a reevaluative epiphany about the legacy of Louis Pasteur. Now hiding in El Oued, an Algerian town, among a handful of Indian emigrants, Alu is reacquainted with Balaram's copy of the *Life of Pasteur* after meeting Mrs. Verma, a microbiologist and, as it happens, the daughter of one of Balaram's fellow Rationalists. Mrs. Verma herself continues the novel's dialogic conceptions of reason in her disillusionment, despite her profession, with scientific rationality. In numerous arguments with the surgeon Dr. Mishra on political subjects such as national development, scientific progress, and rural socialism, Mrs. Verma consistently rejects any programmatic solutions for improving the world. As she says, "Rules, rules. . . . That's how you and your kind have destroyed everything—science, religion, socialism—with your rules and orthodoxies. That's the difference between us: you worry about rules and I worry about being human."[45] These arguments continue when the two clash over cremation rituals after the death of Kulfi, one of Alu's fellow travelers. Dr. Mishra advocates for strict adherence to Hindu custom using holy water, sandalwood, and ghee. Unable to procure the proper items, Mrs. Verma succeeds in substituting carbolic acid, plain wood, and butter against Dr. Mishra's objections. Here, an ironic reversal takes place: the novel links to Western rationality Dr. Mishra's identitarian appeal to pure Hindu custom, implying both to be part of a kind of epistemic continuum, despite the appearance of cultural otherness. Linear, imperialistic reason is thus also depicted as an enchanting force, undoing the binary between pure cultural difference and disenchanting rationality.[46] This enchantment of reason anticipates Ghosh's critique of Munif's nostalgia in "Petrofiction" by showing the continuity between appeals to purity, absolute cultural difference, and the coloniality of Western rationality.

Mrs. Verma's impure cremation, though, in its use of carbolic acid, represents an ironic triumph of Balaram's circular, deterritorialized reason. At the same time, Kulfi's funeral represents the symbolic cremation of Balaram's war on germs—which is to say, his politics of purification. After concluding that "without the germ 'life would be impossible because death would be incomplete,'" Alu adds Balaram's *Life of Pasteur* to the pyre.[47] The novel's resolute embrace of germs also denotes an embrace of messy cosmopolitan entanglements over territorialized global cartographies of difference—which, the novel is at pains to show, supply the imaginative and material foundations of the oil economy. I read this opposition—radical cosmopolitanism against taxonomized globalized difference—as key to the cosmopolitics that Ghosh develops throughout his work.

In conclusion, I want to offer a reevaluation of the novel within Ghosh's oeuvre by discussing the breaking of reason's circle. *The Circle of Reason* remains Ghosh's least mentioned work, most often cited in passing either to positively note its petrocultural and Indian Ocean setting or to bemoan its perceived derivativeness. Stephanie Jones, for example, lays out the critical consensus that negatively compares the novel to Salman Rushdie's *Midnight's Children*, arguing that "against the wild exuberance of Rushdie's writing, Ghosh's book is generally perceived as being overburdened with strange characters and events, and too full of exotic digressions."[48] These stylistic similarities to Rushdie and Latin American magical realism indeed exist in the novel, particularly in how its sweeping picaresque form couples with a propensity for political allegory. It is thus tempting to read the novel as being consistent with what Neil Lazarus might call a kind of Rushdiean "unconscious" defining postcolonial fiction: a celebratory conception of diaspora, which in an abstract sense is said to hold a kind of critical objectivity that Rushdie himself famously described as being "outside the whale."[49]

On the surface, Ghosh's cosmopolitical vision appears to be part of this broad movement. But this reading misses the subtle ways that Ghosh as early as *The Circle of Reason* can be read as responding to and working against this cultural politics. Ghosh is continually interested in how processes of globalization both destroy potentially international solidarities and as a result *produce* forms of nostalgia as likely to be reactionary as resistant from the margins. This can be seen in the narration of the formation of Gulf oil states in the *Circle of Reason*, but other examples abound throughout Ghosh's later work, including in the enclosure of Third Worldist and Non-Aligned solidarities in *In an Antique Land* and the coloniality of liberal free trade ideologies in *The Ibis Trilogy*. In other words, Ghosh never shares the postcolonial theoretical assumption that processes of globalization would necessarily provide the grounds for weakening imagined national boundaries, essentialist appeals to identity, and universalist discourses. For this reason, Ghosh's work reads as especially prescient in our age of rampant ethnonationalist regimes, from Modi's to Trump's, and signals in a larger sense what Indian Ocean studies' particular reading of globalization as enclosure could contribute politically and aesthetically to our historical moment.

Finally, the cosmopolitics present in Ghosh's work, from *The Circle of Reason* to more recent reflections on novelistic form in *The Great Derangement*, can be described as an attempt to retain a kind of humanist optimism without also reproducing an abstract rationalism that narratively conditions our ability (or inability) to represent the multiple human outcomes of (post)colonial modernity. For a novelist who has been described as quasi-Victorian in his fidelity to the form as (in Henry James's famous phrase) a "loose baggy monster," Ghosh consistently questions the ability of the novelistic form to represent the radical relationality of the human experience as it really is—a problem that is now at the forefront of his reflections on the novel and climate change but that nonetheless shares the basic attitude underlying his earlier engagements with oil and the Indian Ocean.[50] One of the ironic prerequisites of this kind of humanism is the discarding of the notion of a singular humanity, which prevents any serious reflection on the putatively species-level subject of the Anthropocene and the colonial ideologies undergirding normative conceptions of the human. But, as *The Circle of Reason* insists, we are still concerned with representation in the double sense that Gayatri Chakravorty Spivak famously noted in "Can the Subaltern Speak?"—not just as an epistemic question but also as a concern for the radical prospects of anticolonial resistance. We might imagine Mrs. Verma admonishing the theoretical conditioning required to accept a conception of humanism that perversely depends on a dehumanized partitioning of human possibilities: "The tyranny of your despotic science forbade you to [say] the one thing that was worth saying; the one thing that was true. And that was: There's nothing wrong with your body—all you have to do to cure yourself is try to be a better human being."[51]

Notes

1. Amitav Ghosh, *The Circle of Reason* (Boston: Mariner Books, 2005), 169.

2. Amitav Ghosh, "Petrofiction: The Oil Encounter and the Novel," *New Republic*, March 2, 1992.

3. Ibid., 30.

4. Isabel Hofmeyr, "Universalizing the Indian Ocean," *PMLA* 125, no. 3 (2010): 723.

5. Bruce Robbins and Pheng Cheah, eds., *Cosmopolitics* (Minneapolis: University of Minnesota Press, 1998).

6. Sugata Bose, *A Hundred Horizons: The Indian Ocean in the Age of Global Empire* (Cambridge, MA: Harvard University Press, 2009), 13.

7. Edward Simpson and Kai Kresse, "Cosmopolitanism Contested: Anthropology and History in the Western Indian Ocean," in *Struggling with History: Islam and Cosmopolitanism in the Western Indian Ocean* (New York: Columbia University Press, 2008), 2.

8. Gaurav Desai, *Commerce with the Universe: Africa, India, and the Afrasian Imagination* (New York: Columbia University Press, 2013), 23.

9. Ngugi wa Thiong'o, *Globalectics: Theory and the Politics of Knowing* (New York: Columbia University Press, 2012), 52.

10. Ghosh, "Petrofiction," 34.

11. Peter Hitchcock, "Oil in an American Imaginary," *New Formations* 69 (Spring 2010): 84.

12. Ibid., 82.

13. Ghosh, "Petrofiction," 30.

14. Ibid.

15. Ibid.

16. Michael K. Walonen, "'The Black and Cruel Demon' and Its Transformations of Space: Toward a Comparative Study of the

World Literature of Oil and Place," *Interdisciplinary Literary Studies* 14, no. 1 (2012): 58.

17. Denise Ferreira Da Silva, *Toward a Global Idea of Race* (Minneapolis: University of Minnesota Press, 2007), 38.

18. Ghosh, "Petrofiction," 30–31.

19. Micheal Angelo Rumore, "The Terror of Translation: Ruins of the Translation in *The Castle of Otranto* and *Vathek*," *Studies in the Fantastic* 3 (Winter 2015 / Spring 2016): 4.

20. Tuomas Huttenan, "Amitav Ghosh's *The Circle of Reason*—Dismantling the Idea of Purity," *Nordic Journal of English Studies* 11, no. 1 (2012): 126.

21. Ghosh, *Circle of Reason*, 99.

22. Ibid., 47.

23. Ibid., 55.

24. Ibid., 56.

25. Ibid., 57.

26. Ibid., 58.

27. Stephanie Jones, "A Novel Genre: Polylingualism and Magical Realism in Amitav Ghosh's 'The Circle of Reason,'" *Bulletin of the School of Oriental and African Studies* 66, no. 3 (2003): 431.

28. Claire Chambers, "Representations of the Oil Encounter in Amitav Ghosh's *The Circle of Reason*," *Journal of Commonwealth Literature* 41, no. 1 (2006): 33.

29. Ghosh, *Circle of Reason*, 244.

30. Ibid., 244–46.

31. Ibid., 252.

32. Ibid., 257.

33. Brian Larkin, *Signal and Noise: Media, Infrastructure, and Urban Culture in Nigeria* (Durham: Duke University Press, 2001), 36.

34. Ghosh, *Circle of Reason*, 257.

35. Ibid., 259.

36. Ibid., 257.

37. Ibid., 262.

38. Ibid., 263–64.

39. Ibid., 261.

40. Michael Watts, "Violent Environments: Petroleum Conflict and the Political Ecology of Rule in the Niger Delta, Nigeria," in *Violent Environments*, ed. Nancy Lee Peluso and Michael Watts (Ithaca: Cornell University Press, 2001), 292–93.

41. Ghosh, "Petrofiction," 30.

42. Ghosh, *Circle of Reason*, 279.

43. Ibid., 285.

44. Reem Alissa, "The Oil Town of Ahmadi Since 1946: From Colonial Town to Nostalgic City," *Comparative Studies of South Asia, Africa and the Middle East* 33, no. 1 (2013): 48.

45. Ghosh, *Circle of Reason*, 409.

46. This notion of reason as an enchanting force draws particular inspiration from Nile Green's important account of the "religious economy of the West Indian Ocean," which challenges the notion that industrialization necessarily disenchants. As Green compellingly demonstrates in the context of Islamic practices circulating through Bombay in the long nineteenth century, the technologies associated with the fossil fuel era—steamships, locomotives, and other "industrially enabled pattern[s] of reproduction and distribution"—elicited more than just reformist political responses. Rather, they also actively produced religiosities "neither uniform in characteristics nor cosmopolitan in outlook, but highly differentiated and parochially communitarian . . . , neither reformed nor modernist, but customary and traditionalist." While Green writes about a period preceding the oil economy depicted by Ghosh in *The Circle of Reason*, the basic point that colonial development in fact multiplies traditionalist claims to identity is a crucial lesson for understanding the troubling ascension of various ethnonationalisms in the neoliberal era, one that Ghosh clearly anticipated. *Bombay Islam: The Religious Economy of the West Indian Ocean, 1840–1915* (Cambridge, UK: Cambridge University Press, 2011), 11.

47. Ghosh, *Circle of Reason*, 396.

48. Jones, "Novel Genre," 433.

49. Neil Lazarus, *The Postcolonial Unconscious* (Cambridge, UK: Cambridge University Press, 2011), 21. See also Salman Rushdie, *Imaginary Homelands: Essays and Criticism, 1981–1991* (London: Granta Books, 1991).

50. Henry James, "From Preface to *The Tragic Muse*," in *The Portable Henry James* (New York: Penguin Books, 2004), 476–78.

51. Ghosh, *Circle of Reason*, 413.

10.

Xerodrome Lube

Cyclonic Geopoetics and Petropolytical War Machines

Simon Ryle

One of the most celebrated recent global petrofictions is Iranian philosopher Reza Negarestani's 2008 novel *Cyclonopedia: Complicity with Anonymous Materials.* Troubling disciplinary boundaries and overloading conventional genre categories, the novel mashes up 1990s Warwick cybernetic geophilosophy with Lovecraftian "weird" horror and a dash of 1970s conspiracy narrative.[1] Negarestani develops an innovative architectonics of the many disparate cultural and material bodies put into circulation by petroleum—desert monotheism, occult Mesopotamian numerology, Western suburbia, and technocapitalist war. *Cyclonopedia* terms the knowledge of these bodies' interconnectivity "paleopetrology." The subject of a devoted 2011 symposium at the New School in New York, of an edited collection of essays (*Leper Creativity*) featuring some of the most celebrated contemporary thinkers and artists,[2] and of a cult online status, *Cyclonopedia*'s hybrid of philosophy and fiction innovates a Capitalocene geopoetics attuned to modernity's wasting of Earth by oil.

Abandoning conventional plotting and characterization, the novel is composed largely of paraphrases of a fictional academic, Dr. Hamid Parsani—a maverick and somewhat paranoid archaeologist of Mesopotamian ruins and ancient Middle Eastern mathematics. This framing technique allows Negarestani to give novelistic dissemination to Parsani's subterranean and conspiratorial theories of oil, which center on the passage, lubricated by oil, toward death. With a proliferating series of neologisms and conceptual innovations, *Cyclonopedia* attempts to name "the logics of petropolitical undercurrents":[3] the "blobjective" interconnections and lubrications supplied to modernity by oil. "A blobjective view," as Negarestani writes, involves a paradigm shift in thinking; it "necessarily diverges from the Earth as a whole towards an entirely different entity, an earth under the process of 'Erathication.'"[4] As an insurgency against the sun's hegemony, which Negarestani terms "Solar Capitalism,"[5] oil stores massive pools of the sun's energy as an underground resistance to solar life. Oil enables and accelerates the pestilences inflicted on the earth by modernity—the desertification process that Negarestani describes as Erathication, whose final goal is the absolute leveling that is the "Tellurian Omega" or "xerodrome."[6]

"Thinking takes place in the relationship of territory and the earth," state Gilles Deleuze and Félix Guattari in describing the geological origins of Greek philosophy.[7] But if the specifically Greek milieu forges Western thought from territorial negotiations of the territories of earth and sea, oil violently reorganizes this grounding of liquidity. In Negarestani's materialist and paranoid philosophy, the Middle East presents an opposed milieu, one in which human cognition has receded in importance and the liquid mass of oil becomes the central protagonist, demonic agency, and lubrication of the xerodrome destination at the (occluded) heart of both petromodernity and Middle Eastern religious ideologies. "Everything belonging to historical time," as McKenzie Wark describes the novel, "is just minor characters."[8] Oil assembles a contemporary "war machine," to appropriate Deleuze and Guattari's term, pivoted on cyclonic insurgent feedback spirals between East and West. Going back to a long-standing biblical struggle, Negarestani terms this relation the "Gog-Magog Axis." In sources as diverse as ancient Zoroastrian doxa, Wahhabism, iconoclasm, and jihadi ideology, Negarestani detects elements of paleopetrology: a desert ideology of resistance to structure, life, and everything erected. These Middle Eastern ideologies, Negarestani writes, "consider everything that is not a desert a violation against the all-consuming hegemony of God. . . . What they actually achieve, and passively cooperate with, is the Tellurian insurgency of the Earth."[9] Petromodernity accelerates monotheistic desert insurgencies that seek to level "all planetary erections" and ultimately to attain "burning immanence with the Sun."[10] Oil lubricates capitalism and technology in an axial synthesis of East and West, "consummating the technocapitalist oecumenon through synthesis with Islamic monotheistic enthusiasm."[11] This situation is reflected in Negarestani's style. Burrowed, emptied, and networked by oil production and distribution, our excavated geology and holey Capitalocene cultural forms both demand a new geopoetics and constitute the historiographic and narrative lubrication of modernity's becoming. Inscribing flows of becoming into the earth's geology, petroleum is a "narration lube,"[12] and its pipelines and oil fields are a "Hidden Writing." In a disfigured subterranean tunneling, the paranoiac neologisms and tentacling geopolitical archaeologies of Negarestani's writing follow the flows of oil, weaving its way between global flows of capital, weapons technology, Abrahamic monotheisms, and raw crude.

Petropolytical Hyperstition

Cyclonopedia traces the scorched ecologies of petromodernity in thrall to oil capitalism. Negarestani describes a hybrid politics of human and inhuman agencies—a "petropolytics," as the novel has it, drawing an etymological reference to the polymer chemicals that materially underpin contemporary political formations. The novel's subtitle, *Complicity with Anonymous Materials*, references this material ungrounding of anthropocentric politics. Middle Eastern monotheistic religions and contemporary oil-based capitalisms are revealed to be the dialectical expressions of a singular evil Tellurian agency, "the black corpse of the sun."[13] Petromodernity accelerates the way this inorganic sentience has for millennia parasitically used humanity as host in order to engineer its own uprising and spread across the face of the earth: "For the middle-eastern countries there is a strategic symbiosis between oil as a parasite and monotheism's burning core."[14] In Negarestani's innovative telling, we moderns are in a "blobjective" state, our subjectivities and cultural orientations complicit with petroleum polymers, because oil flows now underpin all political possibility. Oil, as the novel theorizes with a vocabulary explicitly developed from Deleuze and Guattari's "geophilosophy," both facilitates and demands the ever-accelerating destratifications of territory, geology, ecosystems, tradition, and human potential in order to produce surplus capital that is constantly restratified in the legal and material infrastructures of contemporary war machines and global petro-capitalism.

Due to its complexity, the vast political and ecological totality of oil modernity has not expedited fictive treatment. Oil is perhaps bigger than the conventional novel. Amitav Ghosh has argued that oil to some extent exposes the impotency of fiction in the face of ecocatastrophe and the complex entanglements of late-modern geopolitics.[15] As Stephanie LeMenager summarizes, "It takes a large plot, in fact strategy, to convey the wide-reaching entanglements of oil."[16] Theory-horror, fusing geophilosophy and pulp style, is Negarestani's aesthetic response to this situation. His linguistic coinages and intertwining of theory and fiction seek to grasp the vast, paralyzing entanglements of petropolytics, and for this reason largely eschew traditional novelistic plotting. If oil demands a new mode of writing, *Cyclonopedia*'s horror-philosophy plumbs what that might entail.

Negarestani's oil geopoetics develops the speculative geophilosophical "hyperstition" of the short-lived Cybernetic Culture Research Unit (CCRU). Founded by Sadie Plant and Nick Land at the University of Warwick in 1995, CCRU sought to "libidinize" theory, composing speculative texts with strange neologisms and forging wild combinations of continental philosophy and the visionary futurity of science fiction. CCRU's speculative investigations centered on relations of thought and the earth, which were developed from Deleuze and Guattari's geophilosophy. Geophilosophic speculation was, as Negarestani writes, "capable of developing epistemological germs out of properties of this environment."[17] For Deleuze and Guattari, geography supplies the physical basis for the unfolding of history: "Geography wrests history from the cult of necessity in order to stress the irreducibility of contingency."[18] The history of thought and human becoming, in this development of Nietzsche's materialist philosophy, is bound to contingent interconnections of matter and geological forces. The hyperstition of CCRU was an acceleration of Deleuze and Guattari's method. Combining "superstition" and "hyper," "hyperstition" described for CCRU the way imaginative science fiction and horror texts could, they believed, reverse-engineer new realities that would, "once 'downloaded' into the cultural mainframe, engender apocalyptic feedback cycles."[19]

In texts circulated on early websites in the late 1990s, the CCRU thinkers prophesized a coming "Technomic Singularity"[20] as the end point of the digital networking of ever-accelerating capitalist assemblages of production. Despite their dystopian and apocalyptic prognostications—which were fueled by Thatcherist social disintegration, pioneering internet research platforms, nihilistic philosophy, insomnia, and (probably) poor nutrition[21]—there remains a thrilling residual confidence in the power of writing in many of the CCRU texts. They believed their writing was plugging into the machinic destratifications of the *socius*.[22] More recently, in a nod toward Negarestani's horror, Amy Ireland has placed an emphasis on the occult power of CCRU's method, describing hyperstition as "summoning a demon"[23] into existence. Close to CCRU, Negarestani's writing nevertheless has both more modest and more disarming aspects. If *Cyclonopedia* forms an assemblage in the

Deleuzian sense and paleopetrology describes a mode of hyperstition that is able to exhume ancient and contemporary oil war machines, it is also the case that—as in H. P. Lovecraft's fiction—the insurgent chthonic force therein revealed is fully indifferent to organic life. As with Lovecraft's impotent protagonists, *Cyclonopedia* claims no power but to glimpse the horror of the emergent sun corpse.

Although the Energy Humanities generally avoids rhetoric as giddying as CCRU's, the writers of oil's cultural effects have often felt impelled to neologism. LeMenager describes how fossil capitalism has altered twentieth-century space, creating a "petrotopia": "the now ordinary U.S. landscape of highways, low-density suburbs, strip malls, fast food and gasoline service islands, and shopping centers ringed by parking lots or parking towers."[24] The difficulty of perceiving the strangeness of geopolitical arrangements and armaments, global supply chains, and technologies that enable cheap and readily available oil seems to impel historians and theorists to compound noun constructions, such as Daniel Yergin's concept "Hydrocarbon Society": "It is oil that makes possible where we live, how we live, how we commute to work, how we travel—even where we conduct our courtships."[25] It supplies our food, sustains our lived geographies, and builds our world—and language struggles to name our strange indebtedness to "petromodernity" (another neologism). As the anorganic heart of modernity, oil has reinvented the late-modern subject as "Hydrocarbon Man,"[26] to use another of Yergin's defamiliarizing coinages. As Timothy Mitchell states, "We are learning to think of democracy not in terms of the history of an idea or the emergence of a social movement, but as the assembling of machines."[27] For the Energy Humanities, the dissolution of the subject and political agency—and the attendant need for new vocabularies—constitutes not wild avant-garde or poststructuralist abstraction but a realist attempt to face the contingent material realities of industrial oil modernity.

Exemplifying an earlier hydrocarbon poetics, Jack Kerouac's 1957 novel *On the Road* was born of these shifts. The desire and joy associated with the rapid traverse of space in Kerouac's prose can be seen as an occluded confrontation with the oil dependence of hydrocarbon modernity. Kerouac's is an America at once charted and unknown. The suburbanization of American life during the twentieth century marked a second loss of the land, the first being the disappearance of the foundational American myth of the frontier at the end of the nineteenth century, during the first carbon transition. Various cultural-poetic strategies sought to address this initial traumatic loss, including Emerson's transcendentalism and Whitman's vagabond poetics. The second phase of loss, due to the striated claustrophobia of suburban life, made possible a nostalgic topographics of the American land. The Beat sacralization of speed, such as Kerouac's mythologizing of the Beats' cross-continental automobile journeys and particularly of the heroic driving skills of his central protagonist Dean Moriarty (loosely based on his friend Neal Cassady), positioned automobile speed and the impossible-incessant motion of internal combustion as an answer and escape from the closed and delimited, fully gridded and mapped space of suburban and agricultural America. For Deleuze and Guattari, the beatniks took Henry Miller's repurposing of the city, his "nomadic transit in smooth space,"[28] and produced smooth space from the striated American land. Speed, as Deleuze and Guattari describe, involves a machinic desire quite contrary to movement: "Movement is extensive; speed is intensive." Speed is the cyclonic production of smooth space in their telling: it "constitutes the absolute character of a body whose irreducible parts (atoms) occupy or fill a smooth space in the manner of a vortex."[29] As an intensive vortex, Beat speed sought a deterritorialized experience of cartographic America. The lost America was refound in speed, but in this exhilarating recuperation, the subject and the earth were made beholden to oil capitalism and vulnerable to all manner of oil-based hyperobjects, from massively fracked Alberta tar sands to the climate-disturbed megahurricanes (e.g., Irma, Katrina, and Harvey) of recent years.

Seeking to name the substances underneath and inside our politics and theologies, *Cyclonopedia* innovates a speculative mode of inhuman geopoetics that is responsive to the vortex speed, situation, and scale of oil capitalism. The change from the earlier oil poetics is significant. Whereas Yergin's Hydrocarbon Man and Kerouac's Beat speed register some of the human-level effects of petromodernity, *Cyclonopedia* instead traces the gradual obliteration of the human. A significant forebear, and referenced across the novel, is Deleuze and Guattari's concept of strata from *A Thousand Plateaus*, much of which they attribute to the fictional Professor Challenger (whom they borrow from the science fiction stories of Arthur Conan Doyle). Challenger is an influential stylistic precursor to Negarestani's use of Parsani to frame hyperstitious concepts as fiction. "Strata are acts of capture," Deleuze and Guattari write with Challenger's voice, whereas an "assemblage" is a hybrid construction that joins the strata and inserts itself between them in order to open a relation with total leveling, the absolute desertification that Negarestani terms "xerodrome," or the absence of all strata that Deleuze and Guattari, with Challenger's voice, call "the plane of consistency": "On one side it faces the strata . . . , but the other side faces something else, the body without organs or plane of consistency."[30] "Every mine is a line of flight that is in communication with smooth spaces," Deleuze and Guattari write. "There are parallels today in the problems with oil."[31] Although they have little else to say about oil, their theory of assemblages as productive and combinatory mining machines is a forebear to Negarestani, positioning energy extraction as working toward the total leveling of desert consistency, or zero differentiation. Following Deleuze and Guattari, in *Cyclonopedia*, assemblages such as drills, oil wells, and pipeline infrastructures—but also international organizations such as OPEC and desire-producing fantasies such as the North American imaginary of the automobile—insert themselves between the geological strata that contain oil, burrowing into the strata. They force metabolic destratifications of divided layerings: "The nervous system and the chemistry of war machines smuggled through oil infuse with the western machines feasting on oil unnoticed, as petroleum has already dissolved or refinedly emulsified in them."[32]

Of central influence in Negarestani's updating of Deleuze and Guattari's geophilosophy is CCRU's theory of "geotraumamatics."[33] One of the richest expressions comes in Nick Land's short fiction "Barker Speaks," which describes the core of the earth as a demonic-metallic compound named Cthelll. Land's framing of geotrauma as theory-fiction, delivered via the fictional voice of Professor Barker, serves as a stylistic linkage between Professor Challenger and Negarestani's use of Parsani. A central text of CCRU hyperstition, by the logic of Barker's geotrauma, the unconscious violence that formed and disbanded every geological stratum and era is forever there in Earth's matter, permeating the entire body of the planet. Geological history is trauma, and Cthelll, the burning underworld inner core of the earth, is the planetary unconscious, the repository of this trauma. *Cthelll* is Land's coinage, derived from the noun "chthonic," meaning "subterranean," but also geologizing Lovecraft's subterranean monster Cthulhu: "Trauma is a body . . . , an iron thing . . . , Cthelll: the interior third of terrestrial mass, semifluid metallic ocean. . . . Cthelll is the terrestrial inner nightmare, nocturnal ocean, Xanadu: the anorganic metal-body trauma howl of the earth."[34] As precursor to the demonic oil sentience described by *Cyclonopedia*, Land's speculative theory-fiction charted a revelatory new mode of thinking about the interrelation of creaturely sentience and geological time. Recasting the archaic Carboniferous memories that come to the characters and the earth of speculative sci-fi narratives (such as J. G. Ballard's *The Drowned World*) with giddying and materially repurposed Freudianisms, Land positions the iron cauldron of the earth's center as a repressed memory bank whose convective currents and electromagnetic fields reach tendrils of neurotic violence into all matter and life on Earth. Barker's name aligns his speech with the inhuman (the animal growl) as a suggestion of these deeper (in)organic traces that are recapitulated in more complex life forms. And this means, by the terms of Barker's argument, that all life, at least since the oxygenation of the atmosphere 600 million years ago, carries unconscious imprints of Cthelll in its very being: "Geotrauma is an ongoing process, whose tension is continually expressed—partially frozen—in biological organization."[35] More horrified and paralyzed in the face of human finitude than some other CCRU texts, Land's geotrauma ascribes repressed traumatic memories and sentience to the inner strata of Earth's geology that is recapitulated in all biological life.

If libidinized lexical hyperbole permeated the CCRU texts, then *Cyclonopedia* locates the geological substance consummate to their stylistic excesses. With Negarestani's novel comes the sudden apperception that oil was perhaps always the unconscious central subject of CCRU—though oddly, the Warwick collective seems never to have written about oil. Land's fictional Cthelll reworks geophilosophy as hyperbole, whereas *Cyclonopedia* justifies hyperstition, even making it seem an appropriately measured method, by exposing the irrational excesses of oil: "Petroleum poisons capital with absolute madness."[36] Connecting dystopian anxieties and geophilosophic contingencies, *Cyclonopedia* locates the East-West petro-axis at the heart of the "technocapitalist singularity": "To inquire into the Abrahamic war machine in its relation to the technocapitalist war machine we must first realize which components allow Technocapitalism and Abrahamic monotheism to reciprocate at all. . . . The answer is oil."[37] Turning Deleuze and Guattari's strata and Land's demonic-sentient earth to the specific theologies and petropolytics of oil production, Negarestani overhauls the abstract mechanics of Deleuze and Guattari's model, locating demonic agency in the specific accelerative simultaneity of oil extraction. As a lubricating force, oil sustains the inevitability of its production, and the ever-accelerating Erathication of earth. This means oil extraction as assemblage is also a teleology of destruction, tied by the material dynamics of its own destratifications to the xerodrome plane of consistency, the absolute desertification of earth: "Oil [is] a lubricant, something that eases narration and the whole dynamism toward the desert."[38] Oil is an unraveling fiction of earth trauma, both the writing of history and a kind of cyclonic self-feeding psychosis that keeps on flowing inevitably toward the absolute xerodrome of desert consistency because its flow lubricates its own narrational lubes.

Cyclonopedia in this sense parallels ecocritical concern to name the vast, imperceptible scale of Anthropocene and Capitalocene interventions into the life and mineral systems of the earth. Timothy Morton's concept of the "hyperobject" describes how huge structures are formed out of many interconnected elements. Hyperobjects cannot be seen as such, for they are too complex and large to ever be grasped by individual perception. As Morton details, the late nineteenth century was a great era for hyperobjects, with Marx theorizing capital, Freud the unconscious, and Darwin evolution.[39] None of these hyperobjects could be perceived by an individual. All an individual can perceive of them are small fragments of their effects, symptoms, or manifestations. What Morton does not fully explicate is why it took until the first carbon transition, the era of carbon-driven industrialism via coal-iron-steam-railways, for the hyperobject to emerge into consciousness. As the supply and production networks fueled by coal took hold of modernity, capital, the unconscious, and evolution were each in their own way vital for establishing a new mode of politics that was set against the oppression of workers by capitalists, against the neurotic repression of unconscious desire, and against an anthropocentric worldview that sought to isolate humans from other species.

Morton shows how the political interventions made possible by hyperobjective thinking come even further to the fore in the twentieth century, though again he does not fully acknowledge that it is the second carbon transition (oil-steel-electricity-synthetics-cars-airplanes) that deepens both our awareness of and embeddedness in hyperobjects. This embeddedness is well exemplified by global warming, perhaps the most fearful effect of petromodernity. The ecological strain of the massive scale of carbon industrial production has in recent years raised the level of carbon to an average of 409.92 particles per million in 2018,[40] the highest it has been for the last 20 million years, according to the Intergovernmental Panel on Climate Change.

Yet one might ask: Does knowledge of the oil hyperobject facilitate our resistance to it? As with the viscous pleasure of cruising a freeway listening to My Bloody Valentine (a sensation that Morton links to the hyperobject), arguably both the oceanic oneness and monadic isolation of Morton's "ecological thought" are, rather, cultural products dependent on oil. That there is no part of us (or anything) that is not surrounded and permeated by the oil hyperobject is curiously bound to Morton's ecology. Consider a revealing image: "On every right side mirror of every American car is engraved an ontological slogan that is highly appropriate for our time: OBJECTS IN THE MIRROR ARE CLOSER THAN THEY APPEAR."[41] For Morton, the message on the car mirror describes how we are not the autonomous subjectivities that late modernity might seem to have gifted us but instead are riven by geopolitics. Objects are closer than they appear. Yet, for Morton, this revelation of closeness notably comes from behind the wheel of an automobile. This is an ecocriticism that is "CLOSER" while enclosed within an oil machine. Following Yergin's neologisms and Kerouac's desire for speed, we might say that half a century after *On the Road*'s hydrocarbon poetics, hyperobjectivity describes a "Hydrocarbon Critique." Like Beat speed updated to the Anthropocene, Morton's viscous, oneiric, monadic closeness—cruising the freeways of American oil capitalism to Kevin Shield's oozing guitars—suggests the deterritorializing, oceanic pleasure by which the self enacts its embeddedness within oil assemblages.

This insinuation of oil into contemporary ecologies exemplifies what Negarestani terms "petropolytics." In the paranoid reading of petromodernity's "Hidden Writing" that Negarestani proposes, "the central or main plot is reinvented solely in order that it may stealthily host, transport and nurture other plots."[42] To grasp the totality of oil is not only to give up on a stable sense of human narrative agency, as Morton theorizes, but also to collaborate in Erathication. This is why, rather than filaments of interconnection or closeness, the novel emphasizes infinite decay. Instead of the gentle hyperobjective dissolution of human agency that is always too close and too vast for the subject to perceive, that never actually violates the cognizance of the ecological thinker granted the privileged knowledge of the hyperobject, Negarestani offers the horrified vision of a netherworld demon as the driver of decay—a radically alien sentience that spreads pestilence by interweaving its oily substance with the most intimate aspects of our perceptions, desires, and beliefs.

Always strange, *Cyclonopedia* offers only brief and occluded reference to conventional postcolonial problematics. Rather, Negarestani's poetics mimics the cyclonic Gog-Magog Axis of oil excavation in its destratification and deterritorialization (to borrow Deleuze and Guattari's terms) of the unities of East-West. Although Negarestani's fictional and philosophical references are predominantly Western (in *Cyclonopedia* these are Conrad, Lovecraft, Deleuze and Guattari, and CCRU), they are each made to discourse in astonishing new ways with the Middle East. An embodiment of this is found in the character of Colonel Jackson West, a renegade American who has deserted US forces due to the ineptitude of the American generals in Iraq. A knowing pastiche of both Joseph Conrad's and Marlon Brando's Kurtz, West leads a fearsomely disciplined Delta Force special division, the OBIteration Unit, which launches stealth attacks on Islamists. West is smeared in the media by the Pentagon and pursued by "covert missions underway in Iraq" to court-martial him for treason.[43] As the explicitly Western counterpart of Parsani, West delivers sermons to his devoted men on the "Oily Ethics of War,"[44] remarking, "I'll go to Hell with a can of gasoline in my hand."[45]

Ultimately, Parsani and West converge in their attempts to know oil. Their writings come to bear "striking similarities."[46] East and West alike convulse "along the chthonic stirrings of the blob."[47] Like Parsani, West realizes his profession seethes with the undercurrents of oil:

> West started to search for a way to grasp war as an autonomous entity (to grasp war as a machine with machinic particles and parts). He commenced his expedition from the Mesopotamian necropolis of dust, petroleum and derelict warmachines by exhuming the unfathomably ancient models of grasping-war-as-a-machine.[48]

Each also seems to cipher Negarestani's poetics. Against the upward gush and destructive teleology of the Tellurian Omega, both Parsani and West, as exemplars of *Cyclonopedia*'s geopoetics, burrow into ancient Zoroastrian and Mesopotamian pasts. In so doing, West makes himself into a war machine of oil. Although Parsani is a more ambivalent figure, his resistance is also written by oil. Whichever ancient, occult Middle Eastern cultural practice he is researching, "they all lead to petrological implications."[49] An epistemology of oil's subterranean interconnections, however resistant to xerodrome teleology, also aids its object, "participating in and fathoming Oil as the Tellurian lube of all narrations traversing the Earth's Body." Because "fathoming" the katabatic depths also means burrowing down, opening further the "()hole complex," it is also a mode of participating in and contributing to the "Tellurian lube."[50] In this way, the novel's poetics combines paranoid archaeology with paralyzing horror. Parsani finds constant intimations of petromodernity in ancient Middle Eastern cultures. The "corkscrewing motion" of the Gorgonite *kttk-los*, or *kokloma* (snake spiral, or cyclone), that Parsani uncovers "resembles a drill"[51]—it is to this recurrent cyclonic form of burrowing holey complexes in the earth that the novel's title refers—both diagnosing Parsani's archaeology and seeming to prophesize contemporary oil excavation. The Middle East has long constructed cyclonic war machines of the Tellurian Omega as demonic sentience, as West also comes to perceive. For *Cyclonopedia*, petromodernity means that knowledge comes in spiraling, downward-drilling complicity with oil's ancient chthonic insurgency against the sun. Paleopetrology, in this self-diagnosis, becomes a host of Erathication, and petrofiction becomes just another lube to the coming of xerodrome.

Oil, in the novel's terms, lubricates networks that mediate the rise to the surface of an ancient subterranean death drive. Erathication describes the death drive of the earth's interior that oil brings to the surface: "Earth's dream of this death is realized as the *nigredo* with which terrestrial dreams are so drenched."[52] Erathication involves three precepts: (1) "the levelling of all planetary erections," (2) "the immersion of the planetary body in flows and undercurrents," and (3) "a participation with the Earth as manifest degenerate entity."[53] Oil is, according to the concept of Erathication, the matter and activation of the subterranean thanatopic desire of the earth's core to resist the life it hosts and the solar dependency of life: "Trapping the energy of the sun accumulated in organisms by means of lithologic sedimentation . . . , petroleum is a terrestrial replacement of the onanistic self-indulgence of the Sun or solar capitalism."[54] To sustain these claims, the novel links oil capitalism to a recurrent mythopoetic history of Middle Eastern occultisms centered on the materiality of *nafta*: "This act of submission to the all-erasing desert God is called the religion of *taslim* or submission, that is to say, Islam."[55] Capitalism and Middle Eastern monotheism converge at oil: "Petroleum is able to gather the necessary geo-political undercurrents," Negarestani writes, "required for the process of Erathication or the moving of the Earth's body toward the Tellurian Omega—the utter degradation of the Earth as a Whole."[56]

Unlike Morton's hyperobjects, Negarestani's oil does not just touch on the most intimate details of modernity, ungrounding the strata that sustain meaning; it also has demonic agency. In the novel's horror poetics, it tentacles around human culture and parasitizes its human hosts, rotting structure and life. In the paranoiac-speculative terms of the novel, oil stores massive pools of the sun's energy as an underground insurgency against solar life, and this frees and accelerates a thanatopic agency buried deep in the body of the earth: "Oil as the post-mortem productions of organisms is bound to death."[57] In so doing, Negarestani invokes in his horror an eco-hyperstition of the sentient ecologies theorized in Gregory Bateson's pioneering combinations of anthropology, ecology, and cybernetic theory, which more recently Eduardo Kohn's celebrated ethnography has turned to the collective organic sentience of forest ecologies. In Bateson's information systems ecology, "mind is immanent in the total evolutionary structure."[58] Each sentient individual organism is only a node of a global network: "The individual mind is immanent but not only in the body. It is immanent also in pathways and messages outside the body; and there is a larger Mind of which the individual mind is only a sub-system . . . , immanent in the total interconnected social system and planetary ecology."[59] The vast networked sentience of the reciprocal, interactive ecological system and the interconnected homeostatic evaluations that constantly interact at every level of the network of life make ecosystems massive neural networks. Likewise, Kohn's prizewinning *How Forests Think* presents his ethnographic fieldwork with the Runa people in the Upper Amazon as a revelation of nonhuman thought. Kohn describes the discursive, sentient interactions of Indigenous peoples and forest ecologies, seeking ultimately "to liberate our thinking . . . to open ourselves to those wild living thoughts beyond the human."[60] Paralleling and undergirding these organic mind networks, Negarestani images an inorganic sentience.

Oil constitutes absolute decay, lubricating history with the deathly infestation of oil sentience. With the rapid introduction of the sun corpse as agent at the center of this planetary networking, petromodernity recalibrates the sentient ecosystem as a functioning death drive. Here Negarestani draws from Freud. Seeking to understand the terrible excesses of violence unleashed by World War I, Freud's *Beyond the Pleasure Principle* theorized a death drive at the heart of the conservative impulses in organic matter that sought through the development of the organism to return to the inanimacy prior to life. In a speculative philosophical work written before *Cyclonopedia*, Negarestani uses Freud's mechanics of the death drive to describe and accelerate Marx's analysis of capitalism, in which he finds that capital is impelled to expand into every territory and exhaust every resource. Negarestani describes the logic of capitalism as one of "thanatropic . . . *necrocracy*": "We call this conservative regime of the open system or the organism which forces the dissipation or the thanatropic regression to be in conformity to the dynamic capacity of the organism or the organism's affordable economy of dissipation, *necrocracy*."[61] Recasting capitalism as a *necrocrastic* drive, Negarestani turns Freud's psychological mechanism to the global economic hyperobject. Negarestani is influenced in his reading of Freud by Ray Brassier, for whom all matter shares a longing for entropic dissolution of form. Brassier's *Nihil Unbound* describes a death drive that reaches beyond death, in which inorganic and organic forms are caught in a logic of regression to unbounded formlessness. Yet if the tendency supposedly uncovered by Brassier's theory would flatten all matter to an invariant sameness that seems to leave zero space for political resistance—in a kind of philosophical Erathication—Negarestani's occult treatment of oil capitalism as a uniquely thanatopic matter instead initiates "polytics." On one level, the outlandish occultisms of the novel deny the agency and potency of human politics, but on another, Negarestani places Brassier's entropic insight into discourse with geophilosophy by limiting Brassier's logic of unbounded formlessness to one geological substance and stratum: oil. Speculatively fictionalizing an ancient and subterranean death force buried deep within the earth whose extraction aligns with and is facilitated by capitalist necrocracy, *Cyclonopedia*'s hyperstition darkly mirrors the accelerating death drive of oil capitalism.[62]

By combining a materially specific inorganic death drive with ecologies of nonanthropomorphic sentience, the novel forges an occulted retelling, or Easternization, of geophilosophy as the parasitic decay of the lifeworld. For Deleuze and Guattari, the pipeline is a form of the state's subordinating power, the channeling of destratifying hydraulic force into the "striated and measured, which makes the fluid dependent on the solid."[63] The fluid is determined and directed by state power. This anthropomorphic basis can also be seen in Deleuze and Guattari's assertion that "the war machine was the invention of the nomad, because it is in its essence the constitutive element of smooth space."[64] In their telling, war is essentially human, "not a phenomenon one finds in the universality of nature."[65] For Negarestani, though, the pipeline smuggles the lubrications of the Middle Eastern milieu and its xerodrome war machines into the West. The pipeline and war machines are each inorganic modes of infiltration. The pipeline is an "autonomous vehicle which smuggles Islamic war machines."[66] It may be laid by Western or Islamic state powers; each unwittingly furthers its own dissolution. The struggle is not between the state and the individual, as in Deleuze and Guattari, but between life and the subterranean evil. If ecological systems think in Bateson's and Kohn's cybernetic ethnographies, Negarestani's oil is the carrier of a demonic infestation that accelerates an underworld death impulse toward xerodrome.

The images of tentacles, infestation, and parasitism recurrently feature in the horror poetics by which the novel describes oil's hold on modernity. Like the monstrous Chthulu rising from its evil slumber to ravish the earth in Lovecraft's "weird" tales, oil as Negarestani describes it is a demonic force of death that comes up to ravage the surface. As *Cyclonopedia* self-consciously describes, "Lovecraft's poromechanical universe, or ( )hole complex, is a machine to facilitate the awakening and return of the Old Ones through convoluted compositions of solid and void."[67] Negarestani updates Lovecraft by using tentacles and parasites to describe oil's usurpation of human agency. Wark invokes the figure of the parasite to describe the demonic agency of oil in the novel. Capitalism is the host, and oil is the parasitic organism that transforms capitalism: "The host reinvents the earth as an oil-shitting machine."[68] Benjamin Bratton writes of the parasitoid fungus *Cordyceps unilateralis*, which infects the brain of a species of ant and causes the resulting zombie ants to crawl to the precise height of the jungle canopy that maximizes the fungal spores.[69] As Negarestani writes, "From the gas station to the chthonic oil reservoir via the tentacled edifice of oil pipelines . . . , Bush and Bin Laden are obviously petropolitical puppets convulsing along the chthonic stirrings of the blob."[70] Capital and fundamentalism are made zombie host of the solar corpse, and agency is wrenched from human actants—even purported centers of power—by the lubricating dialectic of oil capital.

This vision of tentacular oil, while outlandish and fantastical, allows Negarestani to pinpoint a vital geopolitical shift in the move between the first carbon transition (coal-iron-steam-railways) and the second (oil-steel-electricity-synthetics-cars-airplanes). The novel's occluded polytics here intersect with the research of Timothy Mitchell on the politics of carbon energy. Unlike coal, which requires many thousands of workers to go deep into the earth to extract it, oil gushes to the surface once drilled. Oil actually wants to come above ground. As Negarestani put it, "There is no equivalent, in oil fields, for the miner."[71] This has a vital bearing on political agency. As Mitchell describes, coal made profound contributions to the political life of many nations by directly facilitating the increasing power of mass democracies at the beginning of the twentieth century:

> The exploitation of coal provided a thermodynamic force whose supply in the nineteenth century began to increase exponentially. Democracy is sometimes described as a consequence of this change. . . . People forged successful political demands by acquiring a power of action from within the new energy system. They assembled themselves into a political machine using its processes of operation.[72]

As economies became dependent on energy flows, democracy flourished because coal miners and their unions came to realize their power. In this new material-territorial milieu, the strike became a powerful technique of disrupting energy flow. Oil workers, though, have little power. Oil does not require nearly so many workers; it flows up spontaneously and thus evades the danger of democratically recalibrating society. As Abdelrahman Munif's *Cities of Salt* details, workers on the first Saudi petroleum plants were unskilled and disposable—they had little ability or understanding of what they were involved in and even less ability to take control of the forces of production.

Liquid flux here gestures at oil's rhizomic powers. Oil furthered liquidity with a technics of flow independent of collective human life. Oil gushes up—driven by underground pressure—from water trapped beneath, from gas trapped above, or sometimes with the assistance of pumps, so "its production require[s] a smaller workforce than coal in relation to the quantity of energy produced."[73] Oil is inhuman labor, as is coal, but oil also distributes itself independently of human labor; it is the sun corpse that gifts itself. One of the reasons for the rapid transition to oil was that oil workers could not interrupt energy flows the same way that coal miners could—this is why Churchill converted the British Navy to oil in 1913, committing the West to a century of military struggles centered on the Middle East in order to evade the democratic advancements of Welsh coal miners. First introduced in Pennsylvania in 1860s, as Mitchell notes, "oil pipelines were invented as a means of reducing the ability of humans to interrupt the flow of energy."[74] Deleuze and Guattari describe a "rhizome network" as a distribution and information system that "may be broken, shattered at a given spot, but . . . will start up again on one of its old lines."[75] The flow lines of oil, in the reckoning of Churchill and the Pennsylvania rail depots, were capitalist rhizomes of relative invulnerability.

Tentacles, parasites, larval infestations, and ideologies of rot as manifestations of oil's demonic sentience are central to *Cyclonopedia*'s ()hole poetics. In their fantastic horror, these agents of decay hone an aesthetics interrogative of the Anthropocene era that many scientists and social theorists now believe is necessary to describe the geological effects of carbon-based capitalism. While it involves an important recognition of ecological destruction, use of the term "Anthropocene" is criticized for erasing difference in its flattening of varying intensities and inputs and thus for casually implicating all humans equally in ecological problems such as global warming, mass species extinction, and nuclear waste. As Andreas Malm points out, the average North American consumes around a thousand times more energy than does the average subsistence farmer in the Sahel. The term equally flattens the way resistance to change comes from a powerful minority. As Naomi Klein writes, "The actions that would give us the best chance of averting catastrophe . . . are threatening to an elite minority that has a stranglehold over our economy."[76]

Although they have differing geophilosophical outlooks, Andreas Malm and Jason Moore each propose the term "Capitalocene" to differentiate the precise human mechanism responsible for contemporary ecological destruction. Yet perhaps "Capitalocene" also does not quite describe the interrelation of oil and war. By every measure, the single greatest contributor to global carbon pollution is the US military. US military spending ($820 billion in 2014) is greater than the combined sums devoted in the United States to education, energy, environment, social services, housing, and new job creation. This includes eight hundred overseas military bases in 150 countries around the world.[77] It has been estimated by NGO Oil Change International that the Iraq War (2003–11) was responsible for more annual carbon emissions than those of 139 nations combined. National security is energy security, and energy security is climate obliteration.

As *Cyclonopedia*'s cyclone poetics stresses, the dominant contemporary geopolitical assemblage is a vortex East-West war machine that is fueled and lubricated by oil. As Negarestani has it, "Oil has undergone a process of weaponization on the Islamic front of [the] War on Terror, and has turned into a fuel for technocapitalist warmachines."[78] With the immense flow of oil westward, to maintain the balance of payments, the West needed to find a very expensive product that it was capable of producing that could be sold to the Middle East in massive quantities. As Mitchell shows, "Arms were particularly suited to this task of financial recycling, for their acquisition was not limited by their usefulness."[79] The international arms trade was necessary to support oil purchases. "The real value of US arms exports more than doubled between 1967 and 1975, with most of the new markets in the Middle East," Mitchell writes.[80] The US economy required war in the Middle East to avoid bankruptcy. Oil could flow out only if war flooded in. This led, as Mitchell writes, to "a US policy of prolonging and exacerbating local conflicts in the Middle East, and on an increasingly disjunctive relationship with the Salafist forms of Islam that had helped defend the mid-twentieth century oil order against nationalist and popular pressures in the region."[81] Instability, war, fundamentalism, arms sales, desertification, and endless war form the assemblage that oil returns to the region from which it is extracted:

> World petropolitics—earth as narrated by oil—in regard to the Middle East and the War on Terror, emerges out of these mutual contaminations between States and desert nomads, facilitated by the holey space of petroleum.[82]

This narrative ungrounding of world politics, lubricated by oil, is the situation both described and replicated in the speculative cyclonic geopoetics of *Cyclonopedia*. Forging a theory-horror consummate to petromodernity, Negarestani opens imagination to the telos of destruction inflicted by the hyperobject of oil capitalism and to the degradation of the imaginary by tentacular petro-epistemologies that parallel the wasting of the earth. Alert to the destratifying war machines that hold thought and the earth in their grasp, Negarestani's innovative aesthetics asks us to face the global processes and assemblages narrated by oil. Rather than human religious or political systems as ungrounded agents free to manipulate the matter of the earth as they choose, the novel positions culture as host to petroleum's parasitical sentience and our zombie-like recapitulation of oil's demands as unwitting complicity. The novel bears horrified witness to the ways that the material logic of oil intersects with politics to lubricate the hybrid, ever-accelerating ()hole-complex through which petromodernity spawns xerodrome.

1. See Nick Land, *Fanged Noumena: Collected Writings 1987–2007*, ed. Robin Mackay and Ray Brassier (Falmouth, UK: Urbanomic, 2011); H. P. Lovecraft, *The Call of Cthulhu and Other Weird Stories*, ed. S. T. Joshi (London: Penguin, 2016); Alan Pakula, dir., *The Parallax View* (Los Angeles: Paramount, 1974); and Sidney Pollack, dir., *Three Days of the Condor* (Los Angeles: Dino De Laurentiis, 1975).

2. Ed Keller, Nicola Masciandaro, and Eugene Thacker, eds., *Leper Creativity* (Brooklyn: Punctum Books, 2012).

3. Reza Negarestani, *Cyclonopedia: Complicity with Anonymous Materials* (Melbourne: re.press, 2009), 16.

4. Ibid., 16.

5. Ibid., 154.

6. Ibid., 18.

7. Gilles Deleuze and Félix Guattari, *What Is Philosophy?*, trans. Hugh Tomlinson and Graham Burchell (New York: Columbia University Press, 1994), 85.

8. McKenzie Wark, "An Inhuman Fiction of Forces," in Keller, Masciandaro, and Thacker, *Leper Creativity*, 42.

9. Negarestani, *Cyclonopedia*, 18.

10. Ibid.

11. Ibid., 16.

12. Ibid., 15.

13. Ibid., 233.

14. Ibid., 32.

15. Amitav Ghosh, "Petrofiction," *New Republic*, March 2, 1992, 29.

16. Stephanie LeMenager, *Living Oil: Petroleum Culture in the American Century* (Oxford: Oxford University Press, 2014), 133.

17. Reza Negarestani, "Notes on the Figure of the Cyclone," in Keller, Masciandaro, and Thacker, *Leper Creativity*, 291.

18. Deleuze and Guattari, *What Is Philosophy?*, 96.

19. Delphi Carstens, "Hyperstition," *o(rphan)d(rift>) Archive*, 2010.

20. Nick Land, "Teleoplexy: Notes on Acceleration," in *#Accelerate: The Accelerationist Reader*, ed. Armen Avanessian and Robin Mackay (Falmouth, UK: Urbanomic, 2014), 519.

21. Simon Reynolds, "Renegade Academia: The Cybernetic Culture Research Unit," Energy Flash, 2009.

22. Robin Mackay and Armen Avanessian, introduction to Avanessian and Mackay, *#Accelerate*, 42.

23. Amy Ireland, "The Poememenon: Form as Occult Technology," Urbanomic, 2017.

24. LeMenager, *Living Oil*, 74.

25. Daniel Yergin, *The Prize: The Epic Quest for Oil, Money and Power* (New York: Simon and Schuster, 1991), 14.

26. Ibid.

27. Timothy Mitchell, *Carbon Democracy: Political Power in the Age of Oil* (London: Verso, 2011), 109.

28. Gilles Deleuze and Félix Guattari, *A Thousand Plateaus: Capitalism and Schizophrenia* (London: Bloomsbury Revelations, 2013), 560.

29. Deleuze and Guattari, *A Thousand Plateaus*, 444.

30. Ibid., 46

31. Ibid., 489.

32. Negarestani, *Cyclonopedia*, 71.

33. Nick Land, "Barker Speaks," in Avanessian and Mackay, *#Accelerate*, 494.

34. Ibid., 498.

35. Ibid., 499.

36. Negarestani, *Cyclonopedia*, 27.

37. Ibid., 16–17.

38. Ibid., 21.

39. Timothy Morton, *Hyperobjects: Philosophy and Ecology After the End of the World* (Minneapolis: University of Minnesota Press, 2013), 47.

40. "Global Carbon Dioxide Growth in 2018 Reached 4th Highest on Record," National Oceanic and Atmospheric Administration, US Department of Commerce, March 22, 2019.

41. Morton, *Hyperobjects*, 27.

42. Negarestani, *Cyclonopedia*, 2.

43. Ibid., 125.

44. Ibid.

45. Negarestani, *Cyclonopedia*, 32.
46. Ibid., 79.
47. Ibid., 20.
48. Ibid., 125.
49. Ibid., 41.
50. Ibid., 42.
51. Ibid., 40.
52. Ibid., 233.
53. Ibid., 16.
54. Ibid., 19.
55. Ibid., 21.
56. Ibid., 17.
57. Ibid., 27.
58. Gregory Bateson, *Steps to an Ecology of Mind: Collected Essays in Anthropology, Psychiatry, Evolution and Epistemology* (Chicago: University of Chicago Press, 1999), 466.
59. Ibid., 467.
60. Eduardo Kohn, *How Forests Think: Towards an Anthropology Beyond the Human* (Oakland: University of California Press, 2013), 228.
61. Reza Negarestani, "Drafting the Inhuman: Conjectures of Capitalism and Organic Necrocracy," in *Speculative Turn: Continental Materialism and Realism*, ed. Levi Bryant, Nick Srnicek, and Graham Harman (Melbourne: re.press, 2011), 192.
62. The implicit ecopoetics of *Cyclonopedia* ought to be mentioned here. A searing critique of the reactionary (and perhaps also deathly) politics of some Latourian New Materialists is to be found in Andreas Malm, *The Progress of This Storm: Nature and Society in a Warming World* (London: Verso, 2018). By limiting Brassier's flat nihilist ontology to oil and responding with horror to the situation of Morton's oozing hyperobjective/hydrocarbon pleasure, Negarestani's "polytics" interrogates the material terrains of potential political struggle that Malm describes. This means if Negarestani's "polytics" implies the impotency of conventional human politics, it must also impel resistance. Aesthetic sensation, claims Deleuze in *Francis Bacon: The Logic of Sensation* (ed. Daniel W. Smith [London: Bloomsbury, 2003]), works directly on the nervous system. Invoking horror at our infiltrated zombie-complicity, *Cyclonopedia* demands, at a bodily level, resistance to xerodrome.
63. Deleuze and Guattari, *A Thousand Plateaus*, 423.
64. Ibid., 485.
65. Ibid., 486.
66. Negarestani, *Cyclonopedia*, 30.
67. Ibid., 44.
68. Wark, "Inhuman Fiction," 42.
69. Benjamin H. Bratton, "Root the Earth: On Peak Oil Apophenia," in Keller, Masciandaro, and Thacker, *Leper Creativity*, 51.
70. Negarestani, *Cyclonopedia*, 20.
71. Ibid., 58.
72. Mitchell, *Carbon Democracy*, 12.
73. Ibid., 36.
74. Ibid.
75. Deleuze and Guattari, *A Thousand Plateaus*, 9.
76. Naomi Klein, *This Changes Everything: Capitalism versus the Climate* (New York: Simon and Schuster, 2014), 18.
77. Nicola Slater, "The US Has Military Bases in 80 Countries. All of Them Must Close," *The Nation*, January 24, 2018.
78. Negarestani, *Cyclonopedia*, 57.
79. Mitchell, *Carbon Democracy*, 155.
80. Ibid.
81. Ibid., 187.
82. Negarestani, *Cyclonopedia*, 57.

11.

Oil Gets Everywhere

Critical Representations of the Petroleum Industry in Spanish American Literature

Scott DeVries

"The jungle has swallowed them!" says the last line of *The Vortex* by José Eustasio Rivera. The exclamation is something more than just an inscription on the gravestone of Arturo Cova and his friends: it is what could also be said about an entire century of Latin American novels: they were swallowed by the mountain, by the plains, by the mines, or by the river.
—Carlos Fuentes, *La nueva novela hispanoamericana*

These lines in an essay by noted Mexican author Carlos Fuentes[1] about the Latin American Boom in Spanish American literature were meant to distinguish the region's earlier texts from what he and fellow authors such as Colombian Gabriel García Márquez, Argentine Julio Cortázar, and Peruvian Mario Vargas Llosa were doing at the time the essay was written. The critical and publishing success of these Boom authors may account for the somewhat dismissive tone in Fuentes's evaluation; the Mexican author seems to have won the day, though—novels set in mountains, plains, mines, and rivers have recently been utterly neglected. Such is the case with a rich subtradition of novels from oil-producing nations such as Mexico, Venezuela, Colombia, and elsewhere. In this chapter, I aim to reconsider and recontextualize two of these now unfamiliar texts: *Huasteca* (1939) by Mexican Gregorio López y Fuentes and *Guachimanes* (Watchmen, 1954) by Venezuelan Gabriel Bracho Montiel.[2]

In a 1992 *New Republic* review of *Cities of Salt* and *The Trench* by Abdelrahman Munif, Amitav Ghosh designates as petrofictions those novelistic representations of the Oil Encounter—that is, of contacts between Indigenous Arabic peoples and representatives of the US petroleum industry in the

Middle East. It was significant to Ghosh's review that Munif was a Saudi dissident born in Jordan and that his novels set in the Arabian Peninsula, *Cities of Salt* and *The Trench*, are compelling stories with Arabian locales and characters conceived as the *objects* of oil exploration and extraction. But despite the fact that "it would be hard to imagine a story that is its equal in drama or in historical resonance," Ghosh observes that "there isn't a Great American Oil Novel" and that "very few people anywhere write about the Oil Encounter."[3] That the novels by Munif are the exception that proves the rule seems to be the view taken by scholars within what has come to be called the Energy Humanities. Imre Szeman, for example, observes "the almost complete absence of oil as subject matter (direct or allegorical) in the literature written during the era when it is dominant,"[4] while Graeme Macdonald suggests that instead of thinking about the "Great *American* Oil Novel," it would be more productive to understand "petrofiction's contemporary identification as a subgenre of literature . . . under the rubric of 'world literature' than it is under that of any national literary corpus."[5] It is the tension between the idea that "petrofictions" have largely gone unwritten and the suggestion that searches under the rubric "world literature" might uncover those that have been written that brings me back to Carlos Fuentes and his infamous denigration of pre-Boom novels.

That most scholars working within the field of the Energy Humanities do not mention the voluminous corpus of Spanish American oil fictions probably owes much to the fact that Fuentes was right about the innovative and groundbreaking quality of Boom literature: nearly all the works by Fuentes, García Márquez, Cortázar, and Vargas Llosa have been best sellers and translated into multiple languages. *One Hundred Years of Solitude* alone has sold over forty-five million copies.[6] The critical and publishing success of Boom authors has meant that nearly all Spanish American oil fictions—written mainly from the 1920s to the 1950s—have been neglected as a result. My aim here, therefore, is to reassess *Huasteca* and *Guachimanes* as a corrective to the idea, likely precipitated in part by Fuentes's seemingly wholesale dismissal of earlier Spanish American literature, that imaginative works about oil have been relatively rare. In the Spanish American context, this has simply not been the case.

The novels by López y Fuentes and Bracho Montiel are part of a long and geographically dispersed literary subgenre that was first published at the beginning of the twentieth century and that has representations of the oil industry from Mexico to Chile and from both the Pacific and Atlantic coasts of South and Central America. I have selected *Huasteca* and *Guachimanes* for their representative qualities, but before I turn to specific moments from these two texts, I give a brief survey to provide a sense of the tradition from which they emerged. In "La suave patria," Mexican poet Ramón López Velarde coined the phrase "los veneros del diablo" (deposits of the devil) for oilfields so construed; this negative image persists in the more than ten Mexican novels and plays centrally featuring oil that were published just before and after expropriation of the industry in 1938.[7] These texts form the immediate literary context for López y Fuentes's *Huasteca*, published in 1939, while more recent texts—including, somewhat surprisingly, *Cabeza de la hidra* (*The Hydra Head*, 1978) by Carlos Fuentes himself—feature an oil industry under the control of the state; less surprisingly, these more recent texts retain the same kind of critical discourse toward the industry as in the earlier novels.

Venezuelan titles include the earliest of all the Spanish American oil fictions: *Lilia* in 1909 by Ramón Ayala. At least twenty other oil-focused fictional texts appeared between the 1930s and the 1980s, including a uniquely Venezuelan kind of text: the petro-dictator novel. The dictator novel belongs to a characteristically Spanish American fictional genre that includes entries by some of the most canonical Spanish American authors, such as Guatemalan Miguel Angel Asturias and Paraguayan Augusto Roa Bastos, along with the aforementioned Vargas Llosa and García Márquez. But in Venezuela, the petro-oriented form of the genre includes at least three titles and owes its existence to the Juan Vicente Gomez regime, in power from 1908 to 1935 when oil exploration and extraction began on a large scale in that nation. Colombia has its own corpus of oil fictions with half a dozen titles, while Ecuadorian, Bolivian, Chilean, and Guatemalan authors have also written novels with oil as a prominent element.

An exhaustive list would be long, yet these texts remained largely regional in reach; they were not best sellers, and they did not persist long in the canon, completely disappearing after Fuentes's essay. But my analysis here will affirm that they are worth remembering for their still-applicable critical representations of the culture, economics, and ecology of oil. In *Huasteca* and *Guachimanes*, the ethical discourse remains starkly, even tragically, relevant today, if perhaps unfortunately overshadowed by that of their literary successors from the Boom. My approach will be to document the ethical, cultural, and aesthetic emphases that nearly unanimously describe the tradition's narrative tendencies as they are manifest in these two novels. Specifically, I consider three things: first, how these texts express unique ethical solutions to problems arising from the extraction, production, sale, and transport of oil; second, how they portray acculturation in the wake of industry development; and third, what aesthetic innovations have been realized as a result of the representation of oil's presence, particularly in the context of communities local to extraction, a circumstance more typical of the Spanish American experience than are issues related to consumption.

Huasteca

In Mexico, the petroliterary production includes the voluminous corpus that I referenced earlier, with most texts having been published before Mexican expropriation had become a political fait accompli in 1938 and several others published later that contain more nostalgic reexaminations of President Lázaro Cárdenas's decision to nationalize the industry. López y Fuentes's *Huasteca*, published one year after Cárdenas's declaration, features the characteristically negative representation of foreign oil company presence in Mexico, but it also includes a final chapter on expropriation. The novel includes ecological, political, and cultural considerations and features a kind of oil aesthetics characteristic of Mexican petrofictions but quite different from many of the country's more established literary traditions.

López y Fuentes's text includes a series of episodes that take place in the Huasteca region, from which the novel takes its title, on the central eastern coast of Mexico near Veracruz. Because of its voluminous and easily drilled petroleum deposits, the region has been referred to as the "Golden Lane" by English-speaking oil executives and includes such sites as the Chapopote seeps, Poza Rica, and Cerro Azul—well #4—considered by some to be the greatest oil well in history, with a daily peak production of 260,858 barrels per day.[8] These oil installations provide the backdrop for the lives of brother and sister Guillermo and Micaela, whose family property lies near where the richly productive oil fields are located. But Guillermo agonizes about whether to sell their land to the big oil companies, and Micaela has the very bad luck to fall in love with and marry Harry, a gambling-addict gringo who wagers away Micaela's wealth at the card table. Both Guillermo's reticence to deal with a locally toxic industry and Micaela's disastrous attachment to Harry are narrative strands that provide the context for conflicts that carry the story forward as they point to the cultural disruption that the oil industry has brought to Mexico. These conflicts center around the effects of a foreign corporate presence on social relationships, perceptions of value, familial connections to inherited lands, conceptions of labor, and ethical and moral priorities, but they are also allegorical of the Mexican nation's ambivalent relationship to a foreign-controlled petroleum industry before nationalization, when the choices were either sell out or embrace alternatives. As it unfolds, *Huasteca* tends much more toward the latter, with critical representations of how the industry abused its access to a Mexican natural resource from the time of the revolution until expropriation became politically expedient in the late 1930s and oil was nationalized.

The novel begins with a gathering at the home of a landowner (the father of Guillermo and Micaela) at around the time the infamous Dos Bocas oil field inferno was about to burn itself out.[9] The host recounts some of what he has heard, most of it highly critical of oil company behavior:

> They say that as many as five miles away from Dos Bocas, the grass doesn't grow, the water is undrinkable, and the livestock has been dying of redwater fever.
>
> I've also heard the story that two engineers had gotten drunk to celebrate the strike and accidentally started the fire when they lit a cigarette; but it's just not true. In fact, they say that those gringos started the whole thing on purpose so the whole world will know how much oil there is here and the company's shares will go up so they can raise capital for new projects.[10]

Here the ecological aspect is paramount, an inextricable function of callous economic calculations. Mathematical calculations reveal that the fallout from this incident involves seventy-nine square miles, or more than fifty thousand acres (about twenty thousand hectares). As Sheena Wilson, Imre Szeman, and Adam Carlson indicate, "Capitalist imperatives ensure that we measure what we value; therefore, transformation will require a radical shift in worldview and how we attribute meaning. Oil prices are indexed daily. That which does not neatly fit into such indexes is rendered valueless as an externality or even casualty to this structural violence: glaciers, clean water, clean air, environmental rights, Indigenous rights."[11] Although his words are not specific to oil, Fernando Mires similarly criticizes growth planning in Latin America as an "economics of the absurd," a characterization that very much fits the Dos Bocas conflagration featured in this moment from *Huasteca*:

> Among "growth economists" there exists a kind of tacit pact that goes far beyond all political convictions. This pact is made evident in the absolute disregard that they show toward any notion that deviates to even the smallest degree from the purely quantitative foundations of the economy that they represent.
>
> "Growth economists," in their conservative and liberal manifestations, because they do not take into account the unquantifiable value of nature in their theories of economic calculation, also cannot properly account for humanity, as humans simply cannot be understood in quantifiable terms.[12]

Both of these observations point to the fact that the seventy-nine square miles of ecological destruction are never accounted for by oil company economists; in fact, we see just the opposite: the fire, in López y Fuentes's account, was intentional, intended to spur positive futures speculation in the company as a means to raise capital. This represents the kind of callous economic maneuvering about which *Huasteca* and the whole Spanish American oil fiction genre has been so indignant.

Another of the more insidious charges leveled against the petroleum industry is for the way oil companies would support (or not support) one side or another in a domestic conflict; this comprises something of the politico-cultural aspect of the critical Spanish American oil fiction discourse. When a corporate oil interest takes a side in a political conflict, the company must ensure that the side it chooses triumphs in the end but also that support can be plausibly denied in case it does not. For example, Francisco Madero's succession to Porfirio Diaz in the presidency has been historically understood as indicative of a more democratic (or at least populist) manifestation of Mexico's political evolution. But López y Fuentes's novel has it as something else entirely:

> The truth, just so you know, . . . is as follows: the tyrant, General Porfirio Diaz, was not overthrown by Madero but by U.S. capitalism; that is, oil companies: the General had adopted a more welcoming attitude toward several British oil companies, and our Good Neighbors to the north, well aware of the economic competition that this would bring, had him overthrown by supporting Madero.[13]

So here, a glorious moment in Mexican history—the overthrow of a dictatorial tyrant—is construed and condemned in *Huasteca* as nothing more than the desire by US companies to keep British petroleum interests out of the Mexican market and thus ensure their own commercial dominance. Mexican politics persisted in this trajectory from 1910—at the beginning of the Mexican Revolution, which was eventually won by forces reconstituted politically as the Institutional Revolutionary Party (PRI)—until 2000, when the PRI finally lost the presidency to the PAN party candidate, Vicente Fox. Thus, the kind of oil company interference that *Huasteca* represents has cast a long shadow in actual Mexican political history.

In other sections of the novel, small holders who do not wish to sell are killed; political campaigns are corrupted; and the companies foment war in order to create more demand for the oil they sell. The representatives of the oil industry who orchestrate these machinations are referred to as "second conquerors," "coyotes," and "buzzards."[14] The negative connotations of this type of descriptive imagery directed toward company thugs is representative of the juncture of aesthetic and political elements in the novel and is typical of other oil fictions throughout Spanish America. But in the peculiarly Mexican political and cultural context represented in *Huasteca*, the aesthetics of the ethical discourse accompany a move toward building the case for nationalization:

> At that time, public opinion began to shift more and more in favor of Mexican workers instead of the oil companies: in favor of workers' demands for better salaries and the right to strike; the public took great interest in the failed negotiations between labor and oil companies and in the subsequent legal battles that went all the way to the Supreme Court. . . .
>
> And just when the companies thought they could continue to bring frivolous lawsuits or abandon them if it was in their economic interest, suddenly the world was shocked by one single word: EXPROPRIATION.[15]

In López y Fuentes's text, as in the actual case, a labor dispute between oil companies and unionized workers results in the decision by Cárdenas to nationalize the industry and expropriate its reserves, facilities, and concessions. This moment is followed in the novel by a variety of reactions: "The People: 'It's a second Declaration of Independence, the real one, economic independence'"; "Someone from the U.S.: 'Mexico, land of the thieves!'"; "An engineer: 'Now these greasers will see how hard it is to run this industry. / 'Where are they going to get the skilled technicians.'"[16] And in the novel's final chapter, a billboard is referenced with its slogan reproduced in all caps:

"OIL EXPROPRIATION MEANS ECONOMIC INDEPENDENCE FOR THE NATION."[17] Such are the aesthetic strategies that accompany the ethical discourse in *Huasteca*: the use of various stylistic forms related to font (unemphasized versus all caps), the incorporation of various textual registers (narrative, dialogue, and a compilation of slogans, insults, and declarations), and an emphasis on the connotative power of language.

Thus, López y Fuentes's novel concludes with a gesture that is both typical of and distinct from the tendencies of the Spanish American petro-novel subgenre. It typifies the tradition in that inherent in the billboard's declaration is the expression of a deeply distrustful and thoroughly critical posture toward foreign oil companies—if independence can be achieved only through expropriation, then corporate control over the industry is revealed as highly undesirable. But it is atypical of other texts, particularly those published in the 1930s from oil-producing non-Mexican Spanish American regions, because nationalization came so much later in those other contexts. The litany of oil company malfeasance in the oil fictions of Venezuela, to which we turn next, is just as numerous and insidious: land theft through swindle, political corruption, ecological contamination, the foment of war, and economic dependence. And just as in the Mexican case, Venezuelan petro-novelists were unanimous in their condemnation of the industry and in their advocacy for alternatives to its control by foreign corporations.

Guachimanes

The nationalization of Mexico's petroleum industry had long been expected because of the rising power of oil workers unions there; as a consequence, US and British companies such as Standard Oil and Royal Dutch Shell had been exploring and developing concessions in Venezuela, where the extraction of oil from the nation's massive reserves has had a far-reaching impact. Miguel Tinker Salas, for example, observes that the petroleum industry

> reached into every corner of the country, directly or indirectly influencing most people's traditional ways of life. . . . As it spurred migration, the industry broke down regional barriers and altered the way Venezuelans defined space and self. . . . Oil also dramatically reshaped the physical environment and existing communities while also giving rise to new settlements. As the industry hired hundreds of workers it

> introduced new patterns of labor relations, notions of time, and housing arrangements within the context of socially and racially stratified labor relations that privileged foreigners.[18]

This characterization of the industry's effects in Venezuela on internal emigration, labor, race relations, and other cultural elements is not exclusive to that country, as the oil fictions from other regions demonstrate. In addition to the list of effects documented by Tinker Salas, however, Spanish American petro-novels feature the representation (and condemnation) of ecological contamination, economic injustice, physical violence, disease, and risk of serious, even fatal, injury. But none so much as those from Venezuela.

I might have made Rómulo Gallegos the focus in this section; his place in the canon of Spanish American authors was cemented by the 1929 publication of *Doña Bárbara* about the painful modernization of the livestock industry on Venezuela's interior plains. When Gallegos turned his attention, in *Sobre la misma tierra*, to the oil industry and its impacts on local cultures around Lake Maracaibo, his status as a noteworthy novelist had already been long established. Even so, the novel has remained among the author's lesser-known works. So I have chosen *Guachimanes* by Bracho Montiel for the way it addresses a wide variety of concerns typical of many other Venezuelan and Spanish American oil fictions but also for the way its narrative discourse advances these cultural critiques and ethical prerogatives in aesthetically unique ways.

In *Guachimanes*, cultural and social alterations, ecological degradation, and other effects that arose in the wake of the development of Venezuela's petroleum industry are articulated in the fictional space of novelistic narrative, but they are documented just as effectively as if the text were nonfiction. In Bracho Montiel's novel, the narrator is implicitly articulated as a defender of what has been aggrieved by the oil industry: labor, traditional cultural values, the environment, prevailing social norms, and the social spaces of local residents. In the first chapter, "Uno menos" (One less), the narrator expresses the ambivalence with which the industry must be viewed: it brings economic benefit but at a great and somewhat hidden cost. In one instance, a local worker is killed when the oil company sends thugs to break up a meeting at which the possibility of forming a union is being discussed; in this case, the narrator uses the first person to express profound ambivalence about an industry's economic benefit in the face of murdered workers:

> I came to see something more pleasant, less cruel. I came to get excited about the hustle and bustle of the great industry that has proven so essential for my country[,] and . . . I kind of feel bothered, unsure, trying to make sense of the contrast between the Americans' nice little houses and the horrible gunshot wound that destroyed a noble heart. "I have to get away," I think. "Get far away from all of this insulting and depressing ugliness. But, how can I avert my eyes from this worker and victim?"[19]

The narrator's reflections upon considering the dead man's body cause him to think that "those eyes were made to never close and their now clouded-over pupils demand something of me. Perhaps some kind of objection. Writing down all that I have seen so far, maybe? I begin to investigate, dig deeper, so as to find out about things that I never could have uncovered if I hadn't decided to literally unearth what lies beneath the land that makes up my country."[20]

These moments seem to offer something of a metafictional counterargument to Ghosh's claim about the lack of novels with oil as a central focus. Not only does *Guachimanes* represent one in a long series of twentieth-century Venezuelan petrofictions, but the narrator engages directly with the desire to "avert one's eyes," or cover up the corruption, violence, and acculturation that lies at the very foundations of the oil industry in Venezuela. Yet instead of doing that, the decision is made to "investigate, dig deeper, and uncover." The result, of course—and this is what makes the moment metafictional—is the novel itself. The narrative documents the often corrupt conditions and circumstances attached to the process of searching for petroleum deposits and securing the title to the land on which wells will be drilled, and it examines how so much of the process can be deadly. Commenting on how these facts have remained largely hidden from public view, Graeme Macdonald finds that "a strongly developed strain of petrocultural theory focuses on the way in which the means and effects of oil are structurally occluded from its mass of consumers, making it less apparent as an *explicit* object in social life and thus a specific topic in and for cultural production."[21] Macdonald is referring here to occlusion as the result of a combination of corporate design, consumer ignorance or apathy, and the general ubiquity of petroleum products without it being obvious that these had their origin in some form of crude. But in the Venezuelan and larger Spanish American contexts, the hidden nature of oil in the cultural production of novels has been more coincidental than anything else; it came from the powerful literary sway of Boom authors such as Fuentes and his literary comrades. So a moment such as the one cited earlier from

Guachimanes now finds double application: the journalist narrator must write because the dead insist on a witness, a witness that must neither be silenced by corporate agents nor forgotten by readers and literary scholars.

In subsequent chapters of *Guachimanes*, there are several other criticisms of industry operations that are also typical of the subgenre as a whole. In chapter 2, "Paisaje negro y bermejo" (Red and black landscapes), ecological fallout from drilling and accidental spills finds a lakeshore "stained by floating oil rainbows" and pipeline infrastructure crossing "the clear waters of rivers forever threatened by spills and leaks until the line gets to the ports. There oil disappears into the bowels of tanker transports, inestimable wealth that is shipped away never to return."[22] Both of these moments describe serious environmental contamination, and the second makes the criticism that threatened waterways constitute the price that must be paid, but for wealth that leaves the country and does not return. Medical negligence is another ongoing theme. In one instance, an employee dies from tetanus poisoning suffered as a result of hands with open sores caused by working on the rig. But the company doctor "will sign a death certificate that never mentions the tetanus microbes that were able to penetrate the small wounds on the worker's calloused hands."[23] This effectively hides a chronic and inevitable condition from future workers who will replace the deceased one.

Additionally, the narrative condemns the company policy of blacklisting potential union organizers, but it also connects this practice to the increased likelihood of worker death. In order for an unskilled laborer to escape the economic curse of the blacklist, he must find ever more dangerous work under an assumed name, as when a worker who has been blacklisted is working on an oil installation when the gas pump explodes. The sand at the site is fused into glass-like projectiles that strike his body at high velocity; the wounds are life threatening. But as the narrative indicates, "If he were a foreigner, a citizen of the U.S., the company would have used a boat equipped with the fastest engine; but he was only a native Venezuelan and it was not a prudent risk to use the speedboat, in a case like this: the motorized skiff was normally reserved only for urgent company emergencies."[24] Thus, the life of a blacklisted and doubly nameless laborer is worth no more than even the slightest risk that the company will suffer economic harm if one of its faster boats is not available for pressing business. The companies have institutionalized and commodified cultural bias to the extent that an abstract principle of economics—the assumption of risk, however small—now represents a threat to profit that is valued higher than the life of an individual Venezuelan. Here we find expression, after the earlier-cited analysis of Wilson, Szeman, and Carlson, of just another anonymous worker who has become a "casualty to the structural violence" of callous corporate economics that assigns no value to externality.

The foregoing illustrates a few of the many other examples of company abuses detailed in *Guachimanes*, but the strident opening passage from the final chapter is perhaps most representative of how a critical ethical discourse can occur in conjunction with aesthetic innovation. This is how that last chapter begins:

> Petroleum! Black fluid from hidden entrails! The saliva of Uncle Sam's gnomes! Lubricant for the hinges and handles of the palatial mansions of the Rockefellers and the Mellons! Elixir of Wall Street! . . . One day you became the symbol of the soul of Venezuela! And like that soul, you dig deep, riotously explode, burn with passion, and inevitably fail.[25]

Here, *Guachimanes* both explicitly gives names ("the Rockefellers and the Mellons") and rather creatively articulates criticism ("fluid from entrails" and "gnome saliva"). The passage references the largely hidden nature of oil in its natural state, usually as deposits buried deep underground. But it also refers to the highly visible and potentially explosive nature of its extraction. As Macdonald has observed about global, or international, petrofiction, it stands in contrast to "the various spaces of the world-oil system [by] registering the experiences of those living and working in those 'concealed' or peripheral zones of extraction."[26] For the oil industry, a burning explosion might have its economic correlate expressed in a profit-and-loss statement but one that is not inclusive of all that has happened at the site. Yet framed as it is in the novel in terms of the soul of Venezuela, such an explosion is far more indicative of the totality of human experience in close proximity to oil blowouts. The passage may seem, at first glance, rather dramatic and somewhat less rigorous than typical economic or sociological analysis, but it nevertheless seems to capture something fundamental about oil, culture, and Venezuelan identity. Such uniquely articulated images elaborate an ethical criticism of the petroleum industry as undesirable for its violence (whether accidental or as the result of force used to impose corporate priorities), for the prospect of inevitable economic failure (usually from the so-called Dutch disease),[27] and for subsequent national disgrace.

These examples from *Huasteca* and *Guachimanes* are typical of the tendency to express an unreserved condemnation of the industry in the Spanish American oil fictions subgenre, especially in the numerous texts from the early to mid-twentieth century when oil really began to dominate the political, social, cultural, and economic spheres in petroleum-exporting nations. Earlier I mentioned the tendency within the Energy Humanities following Ghosh to affirm the absence of oil in literary production. Most scholars in that vein likely had in mind US, British, Canadian, and other English-language literary traditions where a dearth of mid-twentieth-century, explicitly petroleum-industry-focused literary fictions certainly describes the case. But in the Spanish American context, the short stories, novels, plays, and even poetry were so numerous as to have had distinct national iterations in Mexico, Venezuela, and several other Spanish American traditions. This literary corpus claims a robust tradition, perhaps overshadowed by the Boom but just waiting to be rediscovered and brought back to light. It has been my aim in this essay, through the analysis of two representative texts, to do just that.

Notes

1. All quotations in this chapter are from my own translations of the original Spanish.

2. Gregorio López y Fuentes, *Huasteca: Novela mexicana* (Mexico City: Ediciones Botas, 1939); Gabriel Bracho Montiel, *Guachimanes: Doce aguafuertes para ilustrar la novela de petróleo* (Caracas: Fundación Editorial El Perro y la Rana, 2010).

3. Amitav Ghosh, "Petrofiction," *New Republic* 206, no. 9 (March 2, 1992): 30.

4. Imre Szeman, "Literature and Energy Futures," *PMLA* 126, no. 2 (2011): 324.

5. Graeme Macdonald, "Oil and World Literature," *American Book Review* 33, no. 3 (March/April 2012): 31.

6. Alvaro Santana-Acuña, "How *One Hundred Years of Solitude* Became a Classic," *The Atlantic*, May 22, 2017, https://www.theatlantic.com/entertainment/archive/2017/05/one-hundred-years-of-solitude-50-years-later/527118/.

7. Ramón López Velarde, "La suave patria," *Revista el maestro* 3, no. 1 (June 1921): 311.

8. See John Blickwide and Josh Rosenfield, "The Greatest Oil Well in History? The Story of Cerro Azul #4," *Oil Industry History* 11, no. 1 (2010): 191–95.

9. This was an oil blowout and conflagration in 1908 on an oil derrick near Veracruz. According to Shawn William Miller, it could be described as a "veritable volcano, and residents of a town 27 kilometers distant claimed they could read the newspaper at night by its pillar of fire. It burned for two full months and virtually destroyed the forests and fishery upon which the local Huasteca Indians depended. Even after the fire had been extinguished, the oil kept gushing, a complete and total loss of a well that could have produced many millions of barrels of oil over many years." See Miller, *An Environmental History of Latin America* (Cambridge, UK: Cambridge University Press, 2007), 157.

10. López y Fuentes, *Huasteca*, 10.

11. Sheena Wilson, Imre Szeman, and Adam Carlson, "On Petrocultures: Or, Why

We Need to Understand Oil to Understand Everything Else," in *Petrocultures: Oil, Politics, Culture*, ed. Sheena Wilson, Imre Szeman, and Adam Carlson (Montreal: McGill-Queen's University Press, 2017), 11.

12. Fernando Mires, *El discurso de la naturaleza: Ecología y política en América Latina* (San José, CR: DEI, 1990), 67.

13. López y Fuentes, *Huasteca*, 84.

14. Ibid., 56, 60, 217.

15. Ibid., 305.

16. Ibid., 306–8.

17. Ibid., 323.

18. Miguel Tinker Salas, *The Enduring Legacy: Oil, Culture, and Society in Venezuela* (Durham: Duke University Press, 2009), 73.

19. Bracho Montiel, *Guachimanes*, 19.

20. Ibid.

21. Macdonald, "Oil and World Literature," 7 (emphasis mine).

22. Bracho Montiel, *Guachimanes*, 21–22.

23. Ibid., 23.

24. Ibid., 59.

25. Ibid., 83.

26. Graeme Macdonald, "Research Note: The Resources of Culture," *Reviews in Cultural Theory* 4, no. 2 (August 2013): 7.

27. A term coined by *The Economist* in 1977 that describes economic conditions in which development and export of one natural resource within a given national economy means appreciation of that country's currency, making exports more expensive and imports inexpensive. These other export industries collapse, and when a downturn hits the dominant natural resource, the entire economy, because it is reliant solely on that one resource, contracts severely. See "The Dutch Disease," *Economist*, November 26, 1977, 82–83.

12.

Conjectures on World Energy Literature

Imre Szeman

For too long, energy has been treated as a largely neutral input into societies—a necessary element of social life but not one that has any significant, defining impact on its shape, form, and character. The history of modernity, for instance, has been figured in relation to novel developments in literary culture, scientific discovery, the birth of cities, the expansion of individual political freedoms, the structures and strictures of colonialism, and the creation of the global nation-state system. It has seldom been narrated in relation to the massive expansion of socially available energy (and the energy of a specific form: fossil fuels, which are easy to transport and store) and the concurrent redefinition of social practices, behaviors, and beliefs occasioned by this historically unprecedented explosion of access to energy.[1] As we attempt to address the consequences of modern petrocultures and to transition beyond them, we need to include energy in our account of any and every aspect of the social and cultural landscapes we inhabit. Otherwise, we remain stuck, imagining techno-utopian solutions to our environmental crisis instead of getting to work redefining our existing social relations, structures, and behaviors.[2]

What does this mean for the study of literature and of culture more generally? Environmental scientist Vaclav Smil, who has repeatedly drawn attention to the broad historical and social significance of energy, has written that "timeless artistic expressions show no correlation with levels or kinds of energy consumption: the bison in the famous cave paintings of Altamira are not less elegant than Picasso's bulls drawn nearly 15,000 years later."[3] This view of the relationship between energy and cultural production is sustainable only if one imagines the aesthetic as falling completely outside history—as in fact "timeless" in the way suggested here—and is in fact contradicted by Smil's own comments elsewhere in "World History and Energy" about the depth of energy's impact on the development of the forms taken by societies throughout history.[4] Smil's claims about the lack of a relationship between art and energy come in the final section of his essay, in which he cautions against a too simple energy determinism in relation to historical development. One of the significant challenges of introducing energy into cultural and social analysis is figuring out exactly how to plot its impacts and effects. While adding energy to the study of literature and culture might not constitute something akin to a final explanation of these practices and their symbolic significance (as if energy were a substitute for the economic base, in the last instance, determining *both* the character of the economic base and the superstructure connected to it), it is equally problematic to write energy out of the picture as having no significance at all and thus requiring none of our critical and conceptual attention when we engage in the study of literature and culture.

An energy determinism is unsustainable. The claim that energy has no impact on literature and culture is equally unsustainable, whether we understand this impact in a narrowly material sense—in the very substance of the acrylic paints used on modern canvases, the stock used to shoot films, or the electricity required to run printing presses and generate cable signals—or in a social sense, through its figuration of social capacities and expectations.[5] How might one begin to add energy to cultural analysis in a way that captures its force and impact without deferring to either one of these extremes? Does making a link between a specific energy system and a literary period (or movement, or form) open up a new way of analyzing texts? Or does it unnerve not just the how but the *why* of literary studies? Given the significance of energy for societies, we need to begin to add energy to our literary and cultural analysis. But just what does it mean to do so, especially in our accounts and understanding of world literature?

The history of energy is constituted by forms of energy giving way to other forms of energy—for instance, the dominance of water and wind in the United Kingdom in the early nineteenth century moving to coal by the middle to late part of the century.[6] One way it might thus be possible to grasp the impact of energy on literary and cultural production is by engaging in a process of periodization—refiguring literary history around eras defined by wood, tallow, coal, whale oil, gasoline, and atomic power, as in the title of Patricia Yaeger's provocative *PMLA* editor's column on literature and energy.[7]

The aim and intention of Yaeger's loose periodization of energy is to begin a conversation about the material and social impact of energy within literary studies. Such broad energocultural periodizations are rendered immediately problematic by the real character of energy systems. In our own energy period, it can be easy to overstate the degree to which oil or fossil fuels are in fact the dominant form of energy.[8] To begin with, the forms of energy used at any given moment in history are inevitably mixed and changing. As new forms of energy are discovered and added to the mix, energy systems become more and more diverse and complicated; contemporary energy systems are made up of an array of energy forms—hydro, nuclear, solar, wind, and more, in addition to fossil fuels. The introduction of nuclear energy in the period following World War II has done very little to offset levels of fossil fuel use; and while the use of fossil fuels since 1850 has decisively reshaped the sources of energy consumed,[9] the specific mix of the types of fossil fuels used (coal, gas, and oil) shifts and changes year in and year out. In an era that I've come to describe as a "petroculture" (to try to stress the ultimate importance of oil), the use of coal continues to grow, and not just in the developing world.[10]

What might further trouble any easy definition of energy periods are the vast differences in levels and forms of energy use around the world. This is true between developed nations and even more so between developing and developed nations. Canada and Germany are both members of the G8 and are among the world's most developed nations. But the mix of energy forms they use and the level of energy employed per capita makes it difficult to easily figure them as part of a period of single petroculture: Germany has a far greater proportion of sustainable energy than Canada as part of its overall energy portfolio, and it uses about half of the energy per capita that Canada does.[11] When the comparison is made between developed and developing countries, the direct link between access to energy and levels of development becomes all too clear. Per capita energy use in Canada is *twenty-five* times greater than in the Democratic Republic of Congo and close to *twenty* times that in Benin and Haiti. These differences in energy use are connected to histories of colonialism, underdevelopment, and global political and economic power; even in petromodernity, large swaths of the planet are still powered by animal labor and the labor of human bodies. Finally, the assertion of a single petroculture obscures the huge differences in access to power *within* nations. Elites everywhere use more energy per capita than do poorer members of societies. While some elites have no doubt transitioned from fossil fuels to solar power in their homes in wealthier neighborhoods, the use of animal energy, wood, and coal continues to fuel societies around the world.[12]

Despite the complex map of contemporary energy use, it seems to me that energy periodization *can* open up new ways of figuring and analyzing literature in relation to the world in which it is produced. To begin with, a map that shows where energy is used, what kinds of energy are used, how much is used, and who uses it is a map of power, privilege, control, and dispossession. There is a strong correlation between energy use and GDP; increases in energy use per capita inevitably lead to economic growth and higher living standards. The politics of colonialism and postcolonialism were shaped by race and ethnicity, military power, and control over the trade of goods (with colonial powers benefiting at the expense of the colonies they controlled). Colonialism was also a period animated by access to energy and the search for and struggle to control ever more energy. Access to fossil fuel energy—first coal and then oil and gas—was essential to managing and maintaining colonial control; the struggle over energy remains a largely untold story within postcolonial literary studies (the present volume notwithstanding) and in the study of world literature, which has focused on other aspects of the combined and uneven development that has characterized modernity.[13] At a minimum, tracking levels of energy use can allow us to identify nodes of power in the world and to consider the impact of these power differentials on cultural production. Indeed, one could argue that the "world" announced by the category of "world literature" does not in fact come into existence until the beginning of the era of fossil fuels: the production of the imaginary named "world" is fueled by the presence of energy sources, including coal and oil, that make the space of the globe increasingly available and accessible to travel, trade, and political power.[14] A global map of energy, one that also captures differences in energy use within nations, highlights how power is actively materialized, with inevitable impacts on social and cultural capacity and sensibility.

A map of energy power is, however, not unlike our already existing understanding of power and privilege, from colonialism through to the postcolonial, neoliberal present. The strong connection between energy and political power, between levels of per capita energy use and socioeconomic status, means that existing narratives of global power tell us a great deal about the sociocultural impact of energy use, even if the specific importance of energy is rarely named. Might we thus do without critical appeals to energy, assuming that its force and social import will show up in the political, economic, and military structures and forces that it has fueled?

It seems clear to me that this would be a mistake. While it may be that there is a strong correlation between existing narratives of culture and power, the specific role of energy in shaping both cannot just be passed over. Likewise, we should not ignore the specificity of the experiences, social sensibilities, and cultural imaginaries produced by distinct energy systems. There is also a strong political rationale for foregrounding fossil fuels in the analysis of cultural production. Despite the mixed and uneven use of energy around the world at the present time and throughout modernity, it *is* possible to speak meaningfully of a period of petromodernity. Petroleum is the *hegemonic* form of energy at the present time. The claim that something has become hegemonic is intended to capture an organizing principle or a shaping dictate of a period; what is hegemonic about immaterial labor at the present time, for example, is not that the majority of people are involved in such forms of work but that it places demands on all forms of labor to "informationalize, become intelligent, become communicative, become affective."[15] It is in this sense that fossil fuels are hegemonic and that we live in a petroculture. The forms and amounts of energy used by individuals around the world remain mixed, but our cities have been shaped around fossil fuels and not nuclear energy or hydropower. The fantasy of suburban living and the freedom of highways owe nothing to wind farms and solar power; and no country imagines that the way forward in their development is to shape agricultural systems around plow and oxen as opposed to mechanical farming and fertilizers (as environmentally damaging as the latter might be). The enormous, unprecedented energy of fossil fuels has shaped (and continues to shape) our cultural and social imaginary in profound ways, including the character of our political structures and the principles and rationales around which we organize our economic practices and decisions. Being alert to the differences in energy use that I've already pointed to is essential. My intent in claiming that we understand modernity as a *petro*modernity isn't to flatten out these differences in the same way that the Anthropocene has been critiqued for eliminating a more nuanced history of the human impact on the environment.[16] It remains important, however, to identify the dominant rationales or logics that have shaped the world in which these differences exist, including those connected to the forms of energy in use.

In "Conjectures on World Literature," Franco Moretti argues that we need to understand the world literary system as "one, and unequal: *one* literature (*Weltliteratur*, singular, as in Goethe and Marx) . . . but a system which is different from what Goethe and Marx had hoped for, because it is profoundly unequal."[17] So it is too with the system of energy use: one petroculture but a profoundly unequal one. Moretti's essay provides a quick articulation of and justification for his system of "distant reading," which is intended in part to identify dominant logics that have structured the world literary system. For him, the system of *Weltliteratur* is organized by a triangle of forces: "foreign form, local material—*and local form*."[18] Moretti cites the Brazilian literary critic Roberto Schwarz, who provides an example of what the inequality of world literature produces at the level of content: "Foreign debt is as inevitable in Brazilian letters as in any other field. . . . It is not simply an easily dispensable part of the work in which it appears, but a complex feature of it."[19] One might imagine inequalities not only of financial power (which generates debt) but also of energy form and capacity to be figured differently at the center and at the periphery, shaping and forming what Moretti calls "local material[s]" in a strong and profound way. Jennifer Wenzel's "Petro-Magic-Realism" offers an example of literary analysis that foregrounds the impact of energy on literary form at the periphery. In order to offer further specificity and analytic density to the widely used literary category of "magic realism," Wenzel argues that the "political ecology" of Nigerian literature is better described as "petro-magic-realism," "a literary mode that combines the transmogrifying creatures and liminal space of the forest in Yoruba narrative tradition with the monstrous-but-mundane violence of oil exploration and extraction, the state violence that supports it, and the environmental degradation that it causes."[20]

Such mappings of specific encounters between aesthetics and politics via the position of literature within petromodernity are essential; there are all too many oil encounters about which the whole story has yet to be told. If I had to assign a tendency that has emerged in literary critical attention to energy systems, and especially to the petrocultural system that has shaped the colonial encounter and the experience of the modern, it is the attempt to identify the sites and spaces in which the specific configuration of form and content names not only a political tension but also a tension animated by energy differentials. But the inclusion of energy inequality as an axis of analysis is hindered by one major problem—a defining aspect of oil in relation to culture that has to be placed at the center of any conceptualization of energy as an analytic tool within the study of world literature. *The importance of fossil fuels in defining modernity has stood in inverse relationship to their presence in our cultural and social imaginaries*, a fact that comes as a revelatory surprise to almost everyone who engages in critical explorations of energy today. The arts of a world powered by horses and the labor of bodies cannot help but be distinct from the expanded time, space, and power of our own petromodernity. Nevertheless, attention to energy differentials has been largely absent not only from critical investigations of literature and culture but also from literature and culture itself. Those who have begun to grapple with the cultural absence of oil in a period shaped by and around the substance have scanned literary history to find texts that have confronted oil hegemony, with productive results.[21] But questions remain. Why has it been so difficult to locate a literary archive for what now appears to be so important a social and historical force? And despite this absence or gap, why does it seem necessary and important to locate this archive now?

One of the earliest links between energy and aesthetics was made by Amitav Ghosh in his 1992 essay "Petrofiction: The Oil Encounter and the Novel." Ghosh's essay is in part a review of Abdelrahman Munif's monumental *Cities of Salt* quintet, which maps the emergence of American petroleum interests in the Middle East and its world-altering impact on those living on the Arabian Peninsula. "Petrofiction" also offers a broader critique of the failure of literary fiction to address what Ghosh names the "Oil Encounter": the intertwining of the fates of Americans and those living in the Middle East around this commodity, a connection that continues to have far-reaching economic, cultural, and political reverberations. In the twentieth century, US power was strongly tied to oil. For the first half of the century, the United States was the world's first true oil superpower, and it used its energy riches to develop a consumer culture built around automobiles, suburbs, and malls. Given the deep imbrication of the US economy and its politics with petroleum, Ghosh expresses puzzlement that no author has taken up the Oil Encounter as literary subject. Ghosh's essay is now a quarter of a century old, and yet there are still far too few literary fictions that undertake an exploration of what could well be said to have defined the politics of the period since the discovery of oil in the mid-nineteenth century—the struggle over access to oil. For that matter, there are also not very many fictions that have dealt with the social, cultural, and political importance of energy and oil in any way, shape, or form. I'm not suggesting that there are no such fictions—there are, including such one-of-a-kind theory-fictions as Reza Negarestani's *Cyclonopedia: Complicity with Anonymous Materials* (2008) and such recent postcolonial narratives as Helon Habila's *Oil on Water* (2011).[22] Very often, however, the presence of oil in a fiction is little more than shorthand for wealth and power, which could have other sources and be just as well figured in other ways.

Ghosh's insight regarding the absence of oil and the Middle East from modern US fiction has resonated with critics and theorists who have attended to the representational challenge presented by fossil fuels. As the planet's hegemon during the height of petromodernity, we might well expect to find it difficult for US fiction and culture to figure, directly and unfalteringly, as one of the principal sites and sources of the nation's power. From the perspective of the United States, oil is simply what makes the country "go" in a way that doesn't necessitate comment or concern; as a result, it gets lost to the background of the physical apparatus within and against which social, cultural, and political life is played out, no more worthy of comment than furniture or the asphalt covering its streets—indeed, less so.

One of the distinctive elements of oil as an energy source is that, unlike coal energy (which does enter into Victorian-era writing), it is a resource whose consumption is dissociated from its extraction.[23] The historian Christopher F. Jones points out that from the very beginning of the establishment of the oil pipeline system, "the users of oil gained [the] benefit of cheap energy without assuming responsibility for its environmental damage."[24] This ethical dissociation also speaks to oil's place in the cultural and social imaginary—it vanishes to the background, invisible to narrative and thus to critique. What Munif's *Cities of Salt* (1987) captures so perfectly is the blind sense of entitlement of those engaged in extracting oil and the sense of puzzlement of those native to the land whose space and culture have been invaded. Even more powerfully, it shows how the world made by oil can be seen only by those foreign to it. *Cities of Salt* manages to transform oil modernity into the science fictional time and space that it is, occupied by creatures of steel and asphalt animated by the liquid remains of plants millions of years old and whose imperatives and rationales seem so out of joint with the physical environment of the earth that it is hard to believe they are not from elsewhere in the universe. As Munif writes, "What new era had begun—what could they expect of the future? For how long could the men stand it? The night had passed, but what about the nights to come?"[25] The fear and uncertainty that Ibn Rashed and his community experience in *Cities of Salt* is justified. The dissociation that enlivens the Americans is infectious; their appearance in the Middle East to hunt for the substance that fuels them is felt by the Arab community as a rupture in time with unknown consequences.

Munif and other postcolonial writers have managed to grasp some of the social power of oil as a result of imaginative openings that the periphery provides. Just as the colonial system and the mechanisms of postcolonial power are evident to those who have to endure them, one might imagine that the significance of energy for shaping modernity is readily apparent in those spaces where petroculture has yet to take hold. Munif's *Cities of Salt* shows the critical openings that the gap between center and periphery can provide. And yet the representational challenges posed by oil are deeper and more complex than this. Ghosh's interest in Munif's work arises in part because it stands as an exception: it is not just US fiction that fails to name oil as a key element of the human drama of the twentieth century but much of postcolonial fiction as well. For the most part, despite the evident sociohistorical importance of fossil fuels, they have in fact played a very small role in artistic and literary expression during the fossil fuel era. In this case, the exceptions—those texts that have been examined by literary critics working in the Energy Humanities—definitely prove the rule.

I suggested earlier that whether or not it was used everywhere in the world to the same degree (or indeed, whether it was used at all in some places), oil has to be seen as hegemonic—as an energy source that organizes life practice in a more fundamental way than we've ever allowed ourselves to believe. What this means is that despite the absence of fictions that take up oil or the oil encounter directly by making it a part of narrative, form, or both, *any and all* examples of cultural expression in the era of oil have in fact been crucially figured by it. In their introduction to *Oil Culture*, Ross Barrett and Daniel Worden write that "oil culture encompasses the fundamental semiotic processes by which oil is imbued with value within petrocapitalism . . . , the symbolic forms that rearrange daily experience around oil-bound ways of life."[26] It is a point echoed by Frederick Buell in his overview of the framing of cultural and social narratives of the fossil fuel era—for him, "[energy] (especially oil) remains an essential (and, to many, *the* essential) prop underneath humanity's material and symbolic cultures."[27] Might we thus read all of modern literature as a petroliterature and so too make energy a necessary component of literary analysis? The literary critic Graeme Macdonald has offered answers to these questions in his provocative and compelling overview of the key issues that have arisen for critics as they have begun to grapple with the entwined histories of energy and culture. For Macdonald, "fiction, in its various modes, genres, and histories, offers a significant (and relatively untapped) repository for the energy-aware scholar to demonstrate how, through successive epochs, particularly embedded kinds of energy create[d] a predominant (and oftentimes alternative) culture of being and imagining in the world; organizing and enabling a prevalent mode of living, thinking, moving, dwelling, and working."[28] His article unfolds the implications of what fossil fuels have meant for being and imagining in the world and what they have meant, too, for the fictional forms that have developed in relation to this world shot through with "narrative energetics" and "psycho-social dynamics" linked to oil's force and power.

Macdonald argues that the inclusion of energy in literary analysis allows us to better grasp how distinct forms of energy help to make possible distinct modes of being and imagining. It is thus tempting—and the results are interesting enough to try it out—to map specific moments of literature in relation to developments in energy history. Modernism can be read in relation to the dissociation that is a central part of the history of fossil fuel use, identifying in the key dates of 1870 (the moment when more energy was extracted from fossil fuels than from photosynthesis) and 1890 (the year in which more than half of global energy came from fossil fuels) significant moments in the history of energy with resonances in literature and culture. Or one might consider postmodernism in relation to the 1973 OPEC crisis, a moment in which narratives of petromodern futures and US hegemony were deeply unnerved and so, too, the self-certainties of narrative and the Western subject. Then, too, one could read postcolonialism as an affirmation of a new era of colonial power linked to the fact that the moment of decolonization was also one in which the mantle of the biggest producer of fossil fuels was being passed from the United States to Saudi Arabia, Venezuela, and other nations in the Global South. To do so would undeniably enrich the vocabulary of literary and cultural analysis and do so in a manner that forces us to increasingly recognize the significance of fossil fuels in the shape taken by modernity *and* to ask questions about how and why energy has for too long been a "nothing" in our assessments.[29]

But while we might thus enrich the critical literary vocabulary, the addition of energy here hasn't truly added an analytic element that helps us to better understand it. In each of the earlier examples, the work of analysis confers the sociocultural significance of developments in the world of energy by indexing it to changes we already know about; reconfigurations of energy neither produced modernism nor created postmodern style nor brought about the end of colonialism, which is not to say that they were insignificant. The challenge as we begin to figure energy into literary and cultural analysis is to know how to explain the precise social impact of a force whose significance is indexed in part by the fact that its import was not figured in much of the culture produced over an extended period. Macdonald is right: all of modern culture is a petroculture; we might also say that world literature is a world petroliterature. But in identifying world literature as the literature of an oil era, are we doing anything more than asking that we now pay attention to how literature can be used as a space that can be mined for insight into the character of this petro-era, presumably for other purposes—that is, so that we might upend petroculture and bring it to an end?

Ian Baucom has written,

> Although I have for some time accepted the force of Fredric Jameson's dictum that "we cannot, not periodize," until very recently it would not have occurred to me that postcolonial study, critical theory, or the humanities disciplines in general needed to periodize in relation not only to capital but to carbon, not only in modernities and post-modernities but in parts-per-million, not only in dates but in degrees Celsius.[30]

Is it productive or meaningful to undertake such a periodization? One could quibble with Baucom's choice of carbon or temperature as the measure of a new periodization[31] while still understanding the rationale for it—to reimagine literature and culture in relation to environmental concerns. A periodization developed around energy and forms of fuel used captures something that Baucom's proposed periods do not: changes to social capacities and possibilities, to the ways that access to energy—or equally, lack of access to it —creates a predominant way of being and imagining that we usually describe as being modern. Despite all of the caveats, qualifications, and warnings I've provided about reading world literature as a petroliterature—as a literature that has to exist within a system organized around the capacities, fantasies and desires, and imaginative and physical possibilities of oil—reading energy into literature and into the world *does* provide us with critical political resources we might otherwise lack, even if care must be taken with precisely how we figure literature in relation to energy.

There is one final connection between energy and literature that needs to be addressed. I asked at the outset whether including energy in literary studies unnerves not just its how but its *why*. Environmental historians J. R. McNeill and Peter Engelke describe the period since 1945 as the "Great Acceleration," as a period that "unfurled in the context of a fossil fuel energy regime and . . . exponential growth in energy use."[32] It is also, they point out, "the most anomalous and unrepresentative period in the 200,000-year-long-history of relations between our species and the biosphere" and a period that is unlikely to continue.[33] Shouldn't we see literature as a key aspect of this period of Great Acceleration—not only as a practice that gives us insight into the period but also as a category and an activity (writing, reading, and interpreting) that is an outgrowth of fossil fuel use? I'm not suggesting here that literature is synonymous with the fossil fuel era—a claim that comes across as absurd given the historical range of texts studied in any literature department. What I'm proposing is that we think more seriously about the links between the expansion of socially available energy and the coming-into-being of a general social capacity for literary activity—that is, a segment of mass culture that emerged with expanded force and power at the beginning of the twentieth century just as fossil fuels became more abundant (more energy for more people means more culture, a missing piece of the puzzle in Thorstein Veblen's assessment of the rise of mass culture).

There is a largely unarticulated but widely accepted liberal narrative of literature that links it to a developmental account of social capacity: more literature means more narrative, which means more freedom and more social possibilities—as if world literature were, in the end, a policy of the World Bank.[34] One of the chief insights we gain by adding energy to world literature is that it allows us to see that all culture is petroculture because our understanding of culture is linked to oil's initial promise of unending abundance. It is always supposed to be better for there to be ever more culture and ever more energy. In recognition of modern culture's deep ties to energy, the link between literature and energy that critics and theorists have begun to make with ever greater frequency raises questions about the material and ecological weight of literature itself. To date, literature has avoided imagining its relationship to energy through units such as watts/hour or kilojoules/year, seeing itself as a medium that generates no environmental burden even as it proliferates awareness about the catastrophe of climate change. Figuring this metarelationship of literature to energy might well generate an important and original ecological relationship to the apparatuses and objects of modern petroculture and might do so in a manner that would also help undo the damaging self-certainties of the culture of liberal capitalism.

This is an abridged and edited version of the original essay, which can be found in the *Journal of Postcolonial Writing* 53, no. 3 (2017): 277–88.

1. J. R. McNeill and Peter Engelke point out that by 1870, human beings were already using more energy from fossil fuels than the annual amount of energy produced by all the photosynthesis on the planet; since 1860, there have been *one trillion* barrels of oil used. See McNeill and Engelke, *The Great Acceleration: An Environmental History of the Anthropocene Since 1945* (Cambridge, MA: Belknap Press, 2014), 8.

2. Imre Szeman, "System Failure: Oil, Futurity and the Anticipation of Disaster," *South Atlantic Quarterly* 106, no. 4 (2007): 805–23.

3. Vaclav Smil, "World History and Energy," in *Encyclopedia of Energy*, ed. J. Vutler Cleveland, vol. 6 (Amsterdam: Elsevier, 2004), 204, 559.

4. Smil notes that "only a few coastal societies collecting and hunting marine species had sufficiently high and secure energy returns (due to seasonal migrations of fish or whales) such that they were able to live in permanent settlements and devote surplus energy to elaborate rituals and impressive artistic creations (for example, the tall ornate wooden totems of the Indian tribes of the Pacific Northwest)" (550). At a minimum, art as a social practice necessitates "surplus energy," which means that varying levels of surplus energy, in conjunction with broader changes in society animated by shifting forms of energy, would produce distinct modes of art practice as well as altering the social significance of art. In all of the changes that Smil narrates in relation to energy, his identification of art as "timeless" speaks more to his own unwillingness to figure art in relation to energy than to art's apparent ability to sidestep the historical shifts generated by changing forms and levels of energy.

5. See Nadia Bozak, *The Cinematic Footprint: Lights, Camera, Natural Resources* (New Brunswick: Rutgers University Press, 2011).

6. Although there is a tendency to imagine societies as having evolved "naturally" to ever more energy-intensive systems and infrastructures, studies of the transition to fossil fuels highlight the political and social struggles that took place to make this happen. See Bob Johnson, *Carbon Nation: Fossil Fuels in the Making of American Culture* (Lawrence: University Press of Kansas, 2014); Christopher Jones, *Routes of Power: Energy and Modern America* (Cambridge, MA: Harvard University Press, 2014); Andreas Malm, *Fossil Capital: The Rise of Steam Power and the Roots of Global Warming* (New York: Verso, 2016); and Timothy Mitchell, *Carbon Democracy: Political Power in the Age of Oil* (New York: Verso, 2011).

7. Patricia Yaeger et al., "Editor's Column: Literature in the Ages of Wood, Tallow, Coal, Whale Oil, Gasoline, Atomic Power, and Other Energy Sources," *PMLA* 126, no. 2 (2011): 305–26.

8. Christopher F. Jones, "Petromyopia: Oil and the Energy Humanities," *Humanities* 5, no. 2 (2016).

9. Edward Renshaw points out that "animals contributed 52.4 per cent of total work output in the United States in 1850; human workers, 12.6 per cent; wind, water, and fuel wood, 27.8 per cent; and fossil fuels, 6.8 per cent. In 1950, work animals are estimated to have contributed only 0.7 per cent of total work output; human workers, 0.9 per cent; wind, water, and fuel wood, 7.8 per cent; and fossil fuels, 90.8 per cent" (284). "The Substitution of Inanimate Energy for Animal Power," *Journal of Political Economy* 71, no. 3 (1963): 284–92. (Thanks to Jeff Diamanti for bringing this article to my attention.)

10. Christophe Bonneuil and Jean-Baptiste Fressoz point out that "if, in the twentieth century, the use of coal decreased in relation to oil, it remains that its consumption continually grew; and on a global level, there was never a year in which so

much coal was burned as in 2014." *The Shock of the Anthropocene* (New York: Verso, 2015), 101.

11. According to the International Energy Association, in 2013 Canadians used 7,202 kg of oil equivalent per capita; by comparison, Germans used 3,868 kg of oil per capita. The German figure represents a *decrease* from 1971 levels. See World Bank, "Energy Use per Capita," *World Development Indicators*, http://data.worldbank.org/indicator/EG.USE.PCAP.KG.OE.

12. For a recent account of the emergence of a new fuel in Madagascar—charcoal—see Norimitsu Onishi, "Africa's Charcoal Economy Is Cooking. The Trees Are Paying," *New York Times*, June 25, 2016, http://www.nytimes.com/2016/06/26/world/africa/africas-charcoal-economy-is-cooking-the-trees-are-paying.html?_r=0. According to the United Nations Food and Agricultural Organization, over the past twenty years charcoal production in Africa has doubled; today it accounts for 60 percent of total global production. The UN also predicts that demand for charcoal in Africa will double or triple by 2050, in large part as a result of population increases on the continent.

13. Amitav Ghosh, *The Great Derangement: Climate Change and the Unthinkable* (Chicago: University of Chicago Press, 2016).

14. The dates are about right: Goethe claims in 1827 that a world literature is "beginning," while Marx speaks famously in 1848 in the *Communist Manifesto* of a "world literature" arising out of "the many national and local literatures." If these comments predate the oil era, they speak to the capacities and possibilities that are emerging as coal begins to be used to generate an increasingly larger part of the energy used in Europe and the United Kingdom.

15. Nicholas Brown and Imre Szeman, "What Is the Multitude? Questions for Michael Hardt and Antonio Negri," *Cultural Studies* 19, no. 3 (2005): 372–87.

16. See, for instance, Claire Colebrook, "Not Symbiosis, Not Now: Why Anthropogenic Climate Change Is Not Really Human," *Oxford Literary Review* 34, no. 2 (2012): 185–210; and Donna Haraway, "Anthropocene, Capitalocene, Plantationocene, Chthulucene: Making Kin," *Environmental Humanities* 6 (2015): 159–65. Critics of the concept of the Anthropocene are far more common than defenders of it; one exception is Dale Jamieson's argument for an "ethics" that emerges from the concept of the Anthropocene. See *Reason in Dark Time: Why the Struggle Against Climate Change Failed and What It Means for Our Future* (Oxford: Oxford University Press, 2014), 185–92.

17. Franco Moretti, "Conjectures on World Literature," *New Left Review* 1 (2000): 56.

18. Ibid., 65.

19. Ibid., 56.

20. Jennifer Wenzel, "Petro-Magic-Realism: Towards a Political Ecology of Nigerian Literature," *Postcolonial Studies* 9, no. 4 (2006): 456.

21. See Ghosh, *Great Derangement*.

22. In *Cyclonopedia*, oil possesses qualities well beyond those we normally assign to it. Negarastani describes it as nothing less than a "satanic sentience" that "possesses tendencies for mass intoxication on pandemic scales (different from but corresponding to capitalism's voodoo economy and other types of global possession systems)" (Melbourne: re.press, 2008), 26.

23. See Allen MacDuffie, *Victorian Literature, Energy, and the Ecological Imagination* (Cambridge, UK: Cambridge University Press, 2014).

24. Christopher F. Jones, "Petromyopia: Oil and the Energy Humanities," *Humanities* 5, no. 2 (2016): 143.

25. Abdelrahman Munif, *Cities of Salt*, Cities of Salt Trilogy, vol. 1 (New York: Vintage Books, 1987), 222.

26. Ross Barrett and Daniel Worden, eds., *Oil Culture* (Minneapolis: University of Minnesota Press, 2014), xxvi.

27. Frederick Buell, "A Short History of Oil Cultures: Or, the Marriage of Catastrophe and Exuberance," *Journal of American Studies* 46, no. 2 (2012): 270.

28. Graeme Macdonald, "The Resources of Fiction," *Reviews in Cultural Theory* 4, no. 2 (2013): 4.

29. Macdonald writes, "If *all* fiction is potentially energetic, valorizing energy use, then how do we kinetically assert our claims and configure our readings to make it more apparent?"

30. Ian Baucom, "History 4°: Postcolonial Method and Anthropocene Time," *Cambridge Journal of Postcolonial Literary Inquiry* 1, no. 1 (March 2014): 125.

31. "A carbon focus is reductionist, possibly the greatest and most dangerous reductionism of all time: a 150-year history of complex geological, political, economic, and military security issues all reduced to one element—carbon." Thomas Princen, Jack P. Manno, and Pamela L. Martin, "The Problem," in *Ending the Fossil Fuel Era*, ed. Thomas Princen, Jack P. Manno, and Pamela L. Martin (Cambridge, MA: MIT Press, 2015), 6.

32. McNeill and Engelke, *The Great Acceleration*, 2.

33. Ibid., 5.

34. See Amitava Kumar, ed., *World Bank Literature* (Minneapolis: University of Minnesota Press, 2002).

13.

Petrofiction as Stasis in Abdelrahman Munif's *Cities of Salt* and Joseph O'Neill's *Netherland*

Corbin Hiday

To paraphrase Marx: most climate science still dwells in the noiseless atmosphere, where everything takes place on the surface, rather than entering the hidden abode of production, where fossil fuels are actually produced and consumed.
—Andreas Malm, *Fossil Capital*

Through its ubiquitous but mostly invisible existence, oil works to totalize and concretize social relations while simultaneously existing within the world of abstraction and concealment. In this chapter, I propose that we also think of novels as models constructing "social space," in which climate becomes a symptom of the material, extractive production driving petromodernity. When using the term "social space," I have in mind Henri Lefebvre's formulation:

> Though a product to be used, to be consumed, it [social space] is also a means of production; networks of exchange and flows of raw materials and energy fashion space and are determined by it. Thus this means of production, produced as such, cannot be separated either from the productive forces, including technology and knowledge, or from the social division of labour which shapes it, or from the state and the superstructures of society.[1]

I situate such an account of space in relation to petromodernity, illuminating how "property relationships" intersect with "forces of production"[2] and thus necessitate dialectical thinking connected to infrastructural *and* superstructural concerns. My focus here will be the aesthetic engagement with these sociopolitical structures and problematics. In thinking through novelistic productions of spatiality, I hope to move beyond mere mimetic reflection[3] and to recognize the representational challenges posed by fossil fuel extraction. I turn specifically to two novels from two different historical moments—one is in translation and one is interested in cultural translatability; more importantly, though, both were published in the midst of or following crises produced by fossil capital. The two seemingly disparate novels are Abdelrahman Munif's *Cities of Salt* (1989), published on the heels of the 1970s oil crises, and Joseph O'Neill's *Netherland* (2008), published in the midst of the Iraq War and the financial crisis. I argue that both novels are built around a central absence that generates unavoidable disorientation and stasis. With so much fuel, social and spatial relations become entwined. Through attentiveness to formal engagement with spatial relations and production, with metaphors and references to surface and depth, and with presence and absence, we can better understand both *Cities of Salt* and *Netherland* as novels thinking through oil and its spatial contradictions at two discrete historical moments of petromodernity.

Cities of Salt and *Netherland* reach into the subterraneous depths of extraction sites to differently theorize absence and stasis in the age of fossil capital. In these novels, we encounter both expected and unexpected versions of what many have called petrofiction[4] in which the characters grapple with a ubiquitous and totalizing system that relies on its relative invisibility. Whereas most discussions of climate fiction revolve around problems of temporality—a scalar irreconcilability between our present crisis, the deep time of geological carbon formation, and the European Industrial Revolution's large-scale unearthing of this dead matter during the nineteenth century—I attend also to petrofiction's theorization of spatiality, in which the promise of limitless movement results in stasis and paralysis. If, following Lefebvre, social space exists simultaneously as fabricated and determinative—"a product to be used, to be consumed" and "also a means of production; networks of exchange and flows of raw materials and energy fashion space and are determined by it"[5]—then perhaps spatiality under petromodernity functions similarly to the bourgeoisie as discussed by Marx and Engels in *The Communist Manifesto*: the regime of petromodernity "creates a world after its own image."[6] This creation, a mere reproduction and repetition, signals an exhausted modernity in which the promise of progress withers, leaving only destruction and stagnation.

This foundational system of modernity produces its own epistemology—a mode of knowledge that relies on the simultaneous acknowledgment of the indispensable nature of oil, its global scale, *and* its invisibility. In their introduction to *Energy Humanities: An Anthology*, Imre Szeman and Dominic Boyer touch on this absence, noting how "recent critical scholarship has had to account for the ways in which fossil fuels have managed to hide in plain sight/site, evading inclusion in our economic calculations as much as in our literary fictions."[7] It is this last concern, fictions, that this essay takes up; while fiction cannot capture totality through representation, the novel form still takes totality to be its epistemological horizon.[8] Relatedly, and importantly within this historical framework, Stephanie LeMenager refers to the "high point of petromodernity" in relation to both midcentury policy and aesthetic production.[9] In a recent essay, "'This Is the Hell That I've Heard of': Some Dialectical Images in Fossil Fuel Fiction," Andreas Malm ventures into literary criticism through his account of fossil capital, asserting that "a case can be made for reading *all* modern fiction as saturated by fossil fuels, whether it speaks of them or not. But texts that do should command special interest."[10] He suggests widening the archive of what might *count* as fossil fiction while also making a case for a particular attentiveness to texts that help illuminate our contemporary crisis.

With help from thinkers like LeMenager, Malm, and Szeman, I want to pursue the notion of absence in two interrelated ways: thinking of absence as it relates to the larger petrofiction archive and thinking of absence as a trope within particular novels. This twofold absence combined with a particular articulation of spatiality reveals disillusionment and stasis, which are structurally embedded in what we understand today as our petromodernity. Pursuing the intersection between absence and petro-space and the resulting stasis, I argue that in order to conceive of petrofiction in relation to what LeMenager calls our "representational and critical morass,"[11] we must shift our critical terrain from the desire for mimetic fidelity and representational indexicality to the productive formal capacity of literature. I shift my attention now to the question of how a certain multiplicity and assemblage as they relate to oil (the form of oil production) might look like those of the novel.

Munif's *Cities of Salt* presents fiction's most explicit rendering of the oil encounter. Munif stages this encounter as a relation between an unnamed Middle Eastern country (generally understood to be Saudi Arabia) and the United States, and what follows is the fallout from the disillusionment and displacement of Bedouins that Rob Nixon refers to as "the rise of a transnational petromodernity that contained, from the outset, the seeds of its own undoing."[12] If we move from the site of oil extraction to its immaterialized ideological saturation in the realm of speculation and trading, we find O'Neill's *Netherland*, which follows Hans van den Broek, the novel's financier protagonist, through post-9/11 society. The story moves between London and New York City as Hans navigates the loss of a friend and a temporary separation from his wife. While *Cities of Salt* represents a canonical work of petrofiction, I argue that an exploration of *Netherland* expands this archive. Both texts articulate fundamental narrative concerns—such as the problem of particularity and generality, or story and discourse[13]—as they illuminate the tension between events and structure and, most importantly, the *relations* between these levels; relatedly, petromodernity forces us to confront the totalizing system of energy production as individual consumers. According to Malm, the "fossil economy" presents itself to us as a preexisting "socio-ecological structure" in which the "embryonic individual is suspended in its fluid."[14] With this historico-political model in mind, I now turn my attention to Munif's and O'Neill's fictions in order to demonstrate how their works novelistically render these problems of totality and of the sociospatial relations between individual and system.

Cities of Salt and the "Total Paralysis" of Petromodernity

Munif's *Cities of Salt* presents the tension between presence and absence through the character of Miteb, a Bedouin of Wadi al-Uyoun, who is subject to the surveillance of the oil company as he keeps a phantom watch over it. Throughout the novel, his status as living comes into question. One such instance occurs in the description of Miteb's defining obstinacy: "Miteb was now in the shadows, appearing and disappearing, but without anyone taking notice, as if he were no longer there at all, or no longer alive."[15] In one of the novel's many moments of foreshadowing (in this case of the disappearance of Miteb) we see the tension between presence and absence, which seems crucial to Munif's articulation of a politics of anti-imperialism. Miteb and his spectral presence become like the oil he fights against, simultaneously everywhere and nowhere. As Miteb is the novel's quasi-protagonist, the uncertainty about his existence serves to illustrate the disruption of individual lives and entire communities: "This new tribe was a doomed extension of Miteb al-Hathal, of the Atoum and of the life that had been."[16] Miteb stands in for both the individual and the group. Contrary to John Updike's dismissive 1986 review of the novel,[17] this "extension" represents the core of Munif's formal inventiveness, a demonstration of the altered metabolic relations produced by imperial fossil capital; in the novel, this happens between characters and environment and between individual characters and collective organizations—whether Bedouins, American and British petrocrats, or "petro-despots."[18] Unsurprisingly, by the end of the novel, Munif presents Miteb not only as a version of the "return of the repressed," hauntingly situated between particular and general, but also as a collective resistance in the form of a labor strike: "The masses of people moved as one man."[19]

Munif's novel contains a theorization of the interrelations between the concrete and abstract in its construction of a novelistic spatiality that includes particular Bedouin families and communities, but it also locates these particularities within the seemingly atemporal cyclicality of the desert. At stake is the transformation of social space; the imperial imposition of new "productive forces, including technology and knowledge"; and an enforced "social division of labour."[20] As Amitav Ghosh points out in his essay "Petrofiction," the "Arabic title has the connotation of 'the wilderness' or the 'desert,'"[21] further illustrating the type of abstraction and universalizing gesture: although this unnamed country is understood to be Saudi Arabia, it could be any Middle Eastern location, any desert that contains oil below its surface. In *Cities of Salt*, the desert as dominant social space becomes figured as both eternal and mysterious, an object ultimately suited to be shaped, re-formed, and exploited—it is a "product to be used, to be consumed" but "also a means of production."[22] This uneasy tension manifests in how the novel stages the ability (or inability) to truly *know* a place. We see this explicitly in the novel's first paragraph:

> Wadi Al-Uyoun: an outpouring of green amid the harsh, obdurate desert, as if it had burst from within the earth or fallen from the sky. It was nothing like its surroundings, or rather had no connection with them, dazzling you with curiosity and wonder: how had water and greenery burst out in a place like this? But the wonder vanished gradually, giving way to a mysterious respect and contemplation. It was one of those rare cases of nature expressing its genius and willfulness, in defiance of any explanation.[23]

This opening passage, full of rich and suggestive images, also contains contradictions that the novel will pursue in different ways. Most striking here is the autochthonous and autonomous existence of the "green" space within the "harsh, obdurate desert." Not unlike the oil that will later be extracted by the American imperialists, the "water and greenery" either emerges from the earth ("burst[s] out") or falls like manna from the heavens. In either scenario, the important fact remains its difference from "its surroundings." Munif's opening lines build out toward a version of the sublime, beyond comprehension and knowledge of the natural world, "in defiance of any explanation." A sense of wonder, bewilderment, and incomprehension persists throughout the novel, particularly present in the Bedouins' hesitant encounter with the sea and "new things": "They felt afflicted by total paralysis; in this isolated place, which had lost even its name, they were only a band of men besieged, not knowing what to do or what their lives would be like in the days to come."[24] Here, the sublimity from the novel's opening passage has transformed into uncertainty and fear—in a word, they now live under *erasure*.

This absence related to movement and transformation stems from the novel's dialectical unfolding of surplus and lack—an "outpouring of green" and almost too much water in conjunction with a lack of comprehensibility and later an inability to maneuver amid occupation. Rob Nixon points to this disconnect on a characterological level, positioning a surplus of native voices in relation to the dominant yet outnumbered imperialists and petrocrats: "Munif conjures, moreover, a huge chorus of disenfranchised voices, some bewildered, some complicitous, others intrepid in their dissidence, yet all outmaneuvered by American and British imperial forces in league with the oil majors and (if sometimes frictionally so) with the petro-despots too."[25] This "huge chorus of disenfranchised voices" parallels Updike's dismissal but once again functions as a reflection of the novel's scale and the way Munif formally refuses the restrictions of conceiving the novel as solely a container for bourgeois individualism. Instead, *Cities of Salt* contains a series of intertwined and complex sets of relations, all intersecting to illuminate the uneven production of colonial social space, a production illustrative of the unstable grounds of oil fiction.

To return to questions related to the novel's form and theorization of spatiality, I suggest we read Munif's text as formally shifting between narrative surface and subsoil; the readers' footing becomes cyclically and repeatedly subverted, similar to the experiences of the various characters and families. Early in the novel, we are able to see the convergence around erasure and absence, the individual and the collective, with the idea of cyclicality and transformation: "It kept going around in the orbit that made everything, however far from its center, part of it ruled by the same unbending laws."[26] Shortly after this orbital persistence in the face of the changing nature of individuals and the tribe, we see further reflections on change: "The name of Miteb al-Hathal virtually disappeared except in official records. . . . How is it possible for people and places to change so entirely that they lose any connection with what they used to be? Can a man adapt to new times and new places without losing a part of himself?"[27] What appears to be his ghostly apparition reemerges near the end of the novel, further illustrating formal repetition and cycles; but the questions asked in this passage are in many ways the central concerns of the novel and can be extrapolated beyond "a man" (Miteb) in relation to the larger collective of the novel. This novelistic specter also haunts our own changing world in the face of fossil capital's material transformations and alterations that produce climatological effects including the disappearance of places and the forced migration of peoples.[28]

While *Cities of Salt* theorizes a particular spatial imaginary derived from petro-production, it also constructs a future-oriented temporality without dramatizing historical progression as actual progress (in the sense of *Bildung*) derived from Western development. In this regard, the promise of fossil capital's forward march through modernity stalls and can be imagined only as stasis and paralysis combined with destruction and despair. The last sentence of Munif's novel simultaneously provides insight into our current sociopolitical moment of climatological anxiety and destruction and into the unstable status of literary critics within an emergent moment of energy and environmental humanities: "No one can read the future."[29] In gesturing toward an uncertain future beyond its final page, the sentence also marks a particular irony of the novel, as a principal character in the story can in fact read, or imagine, the future: Miteb al-Hathal. Through Miteb, Munif's novel produces not only a type of cyclicality and stasis brought on by Western "development" but also a persistent proleptic imagination. One such future-oriented anxiety appears early on via Miteb's obdurate skepticism: "He sensed something terrible was about to happen. He did not know what it was or when it would happen, and he took no comfort in the explanations offered him from all sides."[30] As thinkers like Rob Nixon and Amitav Ghosh have convincingly demonstrated, one part of our present environmental impasse resides within a failure of the imagination—inadequate thinking necessitates that we imagine (or reimagine) a different future.[31]

Netherland stages similar formal commitments to spatiality through its articulation of a logic of boundlessness and limits and its theorization of the relation between surface and depth. This articulation and theorization function to reveal its engagement with petro-finance. The main character, Hans van den Broek, works as an "equities analyst" for "large-cap oil and gas stocks"[32] after working his "first adult job" at Shell Oil.[33] Hans exists as the literal amalgamation of Shell Oil; he is a Dutchman who lives in London and eventually New York City, and he checks all the necessary boxes in Royal Dutch Shell's multinational and colonial history. Building on these petro-financial narrative details, O'Neill's *Netherland* exists within the shadow of the Iraq War, and Hans's oil-soaked career path provides convenient narrative content that also reveals its formal theorization: a depth hermeneutic that constructs social space along axes of horizontal and vertical imaginaries. *Netherland* produces a tension between conceptions of boundlessness and limits that reveals the ultimate proximity between financialization and the imagined postcolonial social space of cricket.[34]

The novel's financialized framework provides insight into the limitless imagination, represented characterologically by Chuck Ramkissoon, who is infatuated with aphorisms and slogans such as "Think fantastic" and has an unrestricted vision of business: "These things must have limits. But not business. Limits in business are limitations."[35] While he acknowledges the necessity of limits, Chuck's overt patriotism and belief in unregulated, neoliberal market forces leads him to imagine his cricket-related business pursuits without limitations. Cricket thus becomes not only a burgeoning market but also unlimited in its social potential. Chuck looks to immigration trends and population growth in order to imagine potential consumers: "There's almost half a million South Asians in New York alone. . . . You know what they do in their spare time, these kids? They play cricket."[36] For Chuck, the paranoia and unrest of a post-9/11 world that looms throughout *Netherland* can be mended through this colonial sport. As he says, "What's the first thing that happens when Pakistan and India make peace? They play a cricket match. Cricket is instructive, Hans. It has a moral angle. I really believe this. Everybody who plays the game benefits from it. So I say, why not Americans?"[37] In the novel, cricket represents both a potential site of profit for the venturous (and ethically dubious) Chuck and an educational tool, bridging cultures through its particular "moral angle."

The impulse to produce a space that contains sport while possessing limitless financial and moral potential ultimately cannot be realized within the novel. In fact, after Chuck's death (his body is found "in the river with his hands tied up") near the end of the novel, a former business partner reveals to Hans that cricket itself became the limit: "There's a limit to what Americans understand. The limit is cricket."[38] The novel moves between London and New York City, two nodes within the novel's domestic network but, more importantly, also two metropoles representing prominent nodes in the contemporary financial network. These cities not only mediate the banking industry of which Hans is a part but also remind us of the important past and present locales of petro-finance: Standard Oil's corporate headquarters was famously located at 26 Broadway in Lower Manhattan before its dissolution, and today, Royal Dutch Shell maintains offices in London and The Hague.[39]

Netherland constantly returns to and evokes images and language immersed in the tension between surface and depth. In seemingly banal but not accidental references to the "Deutsche Bank Building"[40] in addition to "Baku" and "West Texas,"[41] the novel activates an exploration of an actual netherland, a space below the surface. In addition to explicit and implicit references to oil, the novel contains imagery of a previous energy regime as well, that of King Coal, including "coals of communication,"[42] "parental coal mine," and "a coal-miner's exhaustion and automatism."[43] This produces a vision of the *longue durée* of fossil capital. In a seemingly innocuous domestic scene involving Hans and his son bonding through gardening, the soil becomes illuminating: "When I dug the topsoil, I was taken aback; countless squirming creatures ate and moved and multiplied underfoot. The very ground we stood on was revealed as a kind of ocean, crowded and immeasurable and without light."[44] Hans navigates the subsoil with an explorer's sense of wonder and amazement; the netherland becomes "immeasurable" and a "kind of ocean," not unlike the descriptors often used in reference to oil. Oil not only lurks beneath the surface of our lives, providing the conditions of possibility for almost every conceivable action, but also literally exists under the "very ground" we stand on in the form of pipelines.[45]

Thus, oil looms within the subsoil of *Netherland*. The novel's very title makes unsubtle reference not only to Hans's home country, the Netherlands, but also to the subterranean and subsurface of the earth—a hellish netherworld of hidden extractive production. This is a space contained but "immeasurable." Later on in the novel, the stairwell of Hans's residence, the Chelsea Hotel, becomes claustrophobic and, in Hans's mind, is "hellishly subterraneous."[46] The urban environment becomes metonymically linked to the meditations on depths and is perhaps most fully realized while Hans reflects on New York's underground dead: "I associated this multitude with the vast burying grounds that may be glimpsed from the expressways of Queens, in particular that shabbily crowded graveyard with the monuments and tombs rising, as thousands of motorists are daily made to contemplate, in a necropolitan replica of the Manhattan skyline in the background."[47] Here the depths of New York City are mined, and Hans finds not just an existential history connecting him to the graves but a long colonial legacy and history in which the graves include "Netherlanders and Indians"[48] in addition to the unmarked graves of slaves.

Although Hans's connections to the oil industry exist on the surface of the novel, they remain largely superfluous to its plot, which focuses on Hans's relations with his wife, Rachel, and with Chuck Ramkissoon. At one point in the novel, while having a discussion with Rachel about his friendship with Chuck, she accuses him of neglect: "You don't look beneath the surface."[49] Ultimately, Hans attempts to keep hidden what is below the surface—including the complexities and mysteries of his friend Chuck, the "immeasurable," the "hellishly subterranean," and the "multitude" of urban graves. In Hans's failure to "look beneath the surface," he succumbs to the ideological position imposed by his career in petro-finance, but the novel's theorization of a particular petro-space reveals a whole system of social relations (interactions between surface and depth), in which Hans is only one part. Thus, he is able to "take a guess at the oil production capacity of an American-occupied Iraq and in fact was pressed at work about this issue daily, and stupidly,"[50] but he is "unable to contribute to conversations about the value of international law . . . or the menace of the neo-conservative cabal in the Bush administration, or indeed any of the debates, each apparently vital, that raged everywhere."[51] Hans finds himself stuck between a space of calculative knowledge tied to his employment within the petro-financial industry and an absence of political-ethical commitments in the face of a destructive and controversial war. As Hans himself bluntly puts it, "In short, I was a political-ethical idiot."[52] But Hans's idiocy functions as a manifestation of the totalizing force of petromodernity, presenting itself as the absence of alternatives. For a world in which economic structures are dripping in oil, there is no outside of the dilemmas that Hans faces; he just experiences the contradictions particularly acutely.[53]

Netherland imagines the production of space not just through an attentiveness to depths but also through the technological possibilities of unconstrained boundlessness, paralleling the way Chuck discusses his business pursuits. When Hans finds his marriage and personal life at a standstill, an extension of his earlier problem of the "inability to produce a movement,"[54] he finds some refuge in the virtual production and alteration of geographic space. To supplement his travels to see his young son in London, Hans turns to Google: "There was no movement in my marriage, either; but, flying on Google's satellite function, night after night I surreptitiously traveled to England."[55] Through this technological act, Hans gains control over movement, and through manipulating virtual space, he produces a link between London and New York City. Of course, in many ways *Netherland* hinges on this transnational relationship, whether virtual or actual. At the end of the novel, Hans, back in London, returns to this technology: "I go to Google Maps. It is preset to a satellite image of Europe. I rocket westward, over the dark blue ocean, to America. There is Long Island. In plummeting I overshoot and for the first time in years find myself in Manhattan. It is, necessarily, a bright, clear day. The trees are in leaf. There are cars immobilized all over the streets. Nothing seems to be going on."[56] Again, we encounter immobility, and, tellingly, Hans's desire for mapping space remains consistent with a petro-imperial imaginary. The satellite image takes us from island to island, although in this instance Hans now looks back at New York City from London, and ultimately we return to Chuck Ramkissoon and his unrealized cricket project: "There's Chuck's field. It is brown—the grass has burned—but it is still there. There's no trace of a batting square. The equipment shed is gone. I'm just seeing a field."[57] This moment not only connects Hans back to Chuck but also reveals the failure of cricket to produce an alternative outside of the spatial tension between boundlessness and limits; cricket and its bounded space cannot exist outside of an imperial modernity produced by oil.

In the face of limits and an immobility connected to the false promise of oil's progressive development, the novel attempts to find a solution through domestic resolution. With this plot device, the story attempts to escape the contradictions that arise as oil seeps into the narrative and cricket fails as both homogenizing postcolonial force and neoliberal market experiment. Despite the marital reconciliation between Hans and Rachel (complete with obligatory marriage counseling sessions and getaway to India), Hans and the story are unable to move beyond existing narrative contradictions between surface and depth and between limits and boundlessness; even with this domestic resolution, by the novel's end Hans finds himself where he began—with his son, Jake, and wife, Rachel, explicitly figured here as movement without progress as they sit on the London Eye, circular repetition the only possibility on the giant Ferris wheel. The logic of the narrative and the corresponding petro-financial reality that provides its foundation—below the surface—can be understood only through cyclical movements unable to exist outside of a modernity fueled by oil. The closing scene of the novel reinforces Hans's position in the murky middle between bound and boundless, caught between speculation and "facts" and existing within an uncertain world that is potentially beneficial to the realm of petro-finance but devastating for the countless "numbers of Iraqi dead."[58] But instead of imagining an alternative to this political-ethical dilemma, one very much intertwined with Hans's employment status, the novel ends with spatial uncertainty:

> As a Londoner, I find myself consulted about what we're all seeing. At first, this is easy—there's the NatWest Tower, which now has a different name; there's Tower Bridge. But the higher we go, the less recognizable the city becomes. . . . The difficulty arises from the mishmashing of spatial dimensions, yes, but also from a quantitative attack: the English capital is huge, huge; in every direction, to distant hills—Primrose and Denmark and Lavender, our map tells us—constructions are heaped without respite.[59]

Hans attempts to orient himself and others despite the "mishmashing of spatial dimensions" and "quantitative attack."[60] For Hans, the ability to see and grasp appearances becomes difficult when he is enmeshed within a series of spatial contradictions, exasperated by his proximity to the oil industry, and invested in an ideology that perpetuates invisibility, concealment, appropriation, and limitless growth.

Conclusion

Both *Cities of Salt* and *Netherland* illustrate how the promises of development made possible by petromodernity in fact reveal stasis and immobility as irreparably connected to imperial notions of progress. These novels, understood here to be part of a burgeoning conception of petrofiction, demonstrate this immobility through their theorizations of social space, transformations and figurations of environment and landscape, and shifting social relations. Production within fossil capital shifts from the site of extraction in *Cities of Salt* to an abstracted sphere of financialization in *Netherland*. And in returning to the question of absence and visibility, we see that both novels give us versions of surveillance and the failure to glimpse the energy system in its totality. In *Cities of Salt*, the introduction of a telescope shifts how the emir relates to social life around him in the present and in his imagined future: "His patient, deliberate surveillance made the emir very pensive, and he thought of many times long past and wished he had had the telescope with him then. He told his deputy what an important invention this was and said that someday the mind of man would invent a device using many telescopes, making it possible to see people in faraway places, in Egypt and Syria and even farther away."[61] The emir imagines this technological apparatus will provide a way to *expand* sight beyond Wadi al-Uyoun, but instead it functions only to blur his vision to the reality of imperialist destruction facing the region. Using a similar image, as mentioned earlier, *Netherland* ends with Hans and his family on the London Eye's "celestial circuit,"[62] which functions as an urban panopticon—movement without progress; Hans even admits to "drifting out of the moment" back to New York and memories of his mother, his mind's eye oriented to the past. The novel's final sentence—"Then I turn to look for what it is we're supposed to be seeing"[63]—reminds us of the way oil so often escapes our recognition and how, like Hans, we might not always sufficiently "look beneath the surface."

Notes

1. Henri Lefebvre, *The Production of Social Space*, trans. Donald Nicholson-Smith (New York: Blackwell, 1991), 85.
2. Lefebvre, *Production of Social Space*, 85.
3. In his compelling recent work *The Great Derangement*, Amitav Ghosh asks a series of questions that reveal some crucial limitations to his approach: "What is it about climate change that the mention of it should lead to banishment from the preserves of serious fiction? And what does this tell us about culture writ large and its patterns of evasion?" (Chicago: University of Chicago Press, 2016), 11. First, the tension (or antagonism) between genre fiction and "serious fiction" throughout the book raises questions about the stability of these

categories, and second, Ghosh implies that for a novel to address climate change it must "mention it." This privileging of content and representational mimesis persists throughout *The Great Derangement*.

4. In using this term, I have in mind Imre Szeman's formulation: "We need petrofictions not only in order to narrate the points of encounter between societies and individuals produced by the trade of desirable commodities; we need them because oil (unlike spices) is an ur-commodity: the substance on which the globe depends to heat its homes, to move bodies and goods around, to build and maintain infrastructure—the substance that, for better and for worse, makes the world go round." "Introduction to Focus: Petrofictions," *American Book Review* 33, no. 3 (March/April 2012): 3.

5. Lefebvre, *Production of Social Space*, 85.

6. Karl Marx and Friedrich Engels, *The Communist Manifesto*, trans. Samuel Moore (New York: Penguin Classics, 2002), 224.

7. Imre Szeman and Dominic Boyer, introduction to *Energy Humanities: An Anthology*, ed. Imre Szeman and Dominic Boyer (Baltimore: Johns Hopkins University Press, 2017), 6.

8. For Georg Lukács, "the novel seeks, by giving form, to uncover and construct the concealed totality of life." *The Theory of the Novel*, trans. Anna Bostock (Cambridge, MA: MIT Press, 1971), 60.

9. Stephanie LeMenager, *Living Oil: Petroleum Culture in the American Century* (Oxford: Oxford University Press, 2014), 67.

10. Andreas Malm, "'This Is the Hell That I've Heard Of': Some Dialectical Images in Fossil Fuel Fiction," *Forum for Modern Language Studies* 53, no. 2 (2017): 135.

11. LeMenager, *Living Oil*, 11.

12. Rob Nixon, *Slow Violence and the Environmentalism of the Poor* (Cambridge, MA: Harvard University Press, 2011), 74.

13. In his monograph *Narrative Theory: A Critical Introduction* (Cambridge, UK: Cambridge University Press, 2016), Kent Puckett traces the relation between story and discourse from Aristotle to Marx to Barthes as a fundamental element of how narratives function and think.

14. Andreas Malm, *Fossil Capital: The Rise of Steam Power and the Roots of Global Warming* (New York: Verso, 2016), 12.

15. Abdelrahman Munif, *Cities of Salt*, trans. Peter Theroux (New York: Vintage, 1989), 156.

16. Ibid., 133–34.

17. Updike criticized *Cities of Salt* for its interest in "men in the aggregate" as opposed to "individual moral adventure." Quoted in Ghosh, *Great Derangement*, 77.

18. Nixon, *Slow Violence*, 74.

19. Munif, *Cities of Salt*, 613.

20. Lefebvre, *Production of Social Space*, 85.

21. Amitav Ghosh, "Petrofiction: The Oil Encounter and the Novel," *New Republic* 206, no. 9 (1992): 31.

22. Lefebvre, *Production of Social Space*, 85.

23. Munif, *Cities of Salt*, 1.

24. Ibid., 191.

25. Nixon, *Slow Violence*, 74.

26. Munif, *Cities of Salt*, 133–34.

27. Ibid., 134.

28. For a discussion on the importance of an environmental perspective that includes postcolonial thinking in relation to imperial capitalism and the refugee crisis, see Naomi Klein's essay "Let Them Drown: The Violence of Othering in a Warming World," *London Review of Books* 38, no. 11, June 2, 2016, https://www.lrb.co.uk/v38/n11/naomi-klein/let-them-drown.

29. Munif, *Cities of Salt*, 627.

30. Ibid., 31.

31. Ghosh writes, "Indeed, this is perhaps the most important question ever to confront *culture* in the broadest sense—for let us make no mistake: the climate crisis is also a crisis of culture, and thus of the imagination." *Great Derangement*, 9.

32. Joseph O'Neill, *Netherland* (New York: Vintage Books, 2008), 25–26.

33. Ibid., 44.

34. For a reflection on cricket and the colonial experience, see C. L. R. James, *Beyond a Boundary* (Durham: Duke

University Press, 2013). As we learn throughout the course of the novel, Chuck Ramkissoon, like James, was born and spent his youth in Trinidad, perhaps not coincidentally a country that derives most of its wealth from the production of petroleum, petrochemicals, and natural gas.

35. O'Neill, *Netherland*, 162.

36. Ibid., 79.

37. Ibid., 211.

38. Ibid., 230, 251.

39. Here I am reliant on the history by Daniel Yergin, *The Prize: The Epic Quest for Oil, Money and Power* (New York: Free Press, 2009), 31, 765.

40. O'Neill, *Netherland*, 78. As Rachel Havrelock notes, Deutsche Bank's history runs through oil: "The foundational territorial form of the modern Middle East reflected European petroleum dreams; recalibrating concessions factored among the goals of this theater of war. In the prewar oil agreement of 1914, Ottoman allied Germany secured a 25% share for Deutsche Bank along with Royal-Dutch Shell and British investors. The war prevented any further steps to secure the concession, yet England and France built the assumption of German shares into the Sykes-Picot Agreement." See "1917: Oil and the Origins of Middle Eastern Sovereignty [in Hebrew]," *Theory and Criticism*, no. 49 (Winter 2017): 265.

41. O'Neill, *Netherland*, 180.

42. Ibid., 139.

43. Ibid., 93.

44. Ibid., 63.

45. Here I have in mind Imre Szeman's "Introduction: Pipeline Politics," *South Atlantic Quarterly* 116, no. 2 (April 2017): 402–7.

46. O'Neill, *Netherland*, 110.

47. Ibid., 155.

48. Ibid., 239.

49. Ibid., 166.

50. Ibid., 99.

51. Ibid., 100.

52. Ibid.

53. As Timothy Mitchell notes in *Carbon Democracy*, "the economy" as an "object of calculation and a means of governing populations" emerged during the "mid-twentieth century" and "was made possible by oil, for the availability of abundant, low-cost energy allowed economists to abandon earlier concerns with the exhaustion of natural resources and represent material life instead as a system of monetary circulation—a circulation that could expand indefinitely without any problem of physical limits." *Carbon Democracy: Political Power in the Age of Oil* (New York: Verso, 2011), 234.

54. O'Neill, *Netherland*, 99.

55. Ibid., 123.

56. Ibid., 252.

57. Ibid.

58. Ibid., 181.

59. Ibid., 254.

60. Ibid.

61. Munif, *Cities of Salt*, 408.

62. Ibid., 254.

63. Ibid., 256.

Memoirs and Interviews

14.

Assessing the Veracity of the Gulf Dreams

An Interview with Author Benyamin

Maya Vinai

Introduction

Over the last few decades, Indian English authors have written extensively about the problematics of the diasporic community situated in the West. Writers of repute such as Rohinton Mistry, Kiran Desai, Bharathi Mukherjee, and Jhumpa Lahiri, along with many others, have debated vociferously on issues like identity, belonging, space, home, and nation. But there has been a tremendous dearth of literature about the experiences and encounters of South Asian migrants to Gulf countries. Benyamin (Benny Daniel) is one of India's most popular writers who focuses on the life of Malayalees living in the Gulf. His life and experiences in the Kingdom of Bahrain since 1992 have given him much insight into the lives of Gulf migrants. *Goat Days* (2008; originally written in Malayalam and translated into English by Joseph Koiyapally) is one of the earliest Indian novels that re-creates the so-called Gulf experience of migrant laborers. The book was the recipient of a 2009 Sahitya Academy Award and was short-listed for the 2013 DSC Prize for South Asian Literature. Recently, the one hundredth edition of the book was published. Benyamin's book *Jasmine Days* (2014) was the winner of the 2018 JCB Prize for Literature.

Goat Days is set in Kerala, at the southernmost tip of India. It is the first-hand narrative of a sand miner named Najeeb Ahmad, who travels to Riyadh via Mumbai with the dream of earning petrodollars. Unaware of the dangers that lie ahead, Najeeb becomes completely disillusioned when he faces the stark reality of Gulf life, which includes inhuman working conditions, a lack of camaraderie, and the extremely hostile climate. As the plot develops, the readers witness Najeeb escaping from his cruel manager, who is referred to in the novel as the arbab. The arbab not only victimizes workers like him but also confiscates his passport, thereby eliminating any possible chances for escape. *Goat Days* brilliantly captures Najeeb's psychological turmoil while he attempts to retain his sanity in the most inhumane and adverse circumstances and while he and two accomplices attempt to return home. An engineer by profession, Benyamin has to his credit seven novels, three short-story collections, and one travelogue. This interview focuses on key issues such as the plight of migrant laborers in Gulf countries, narrative strategies employed in *Goat Days*, the role of women popularly known back home as Gulf wives, and the response of the host countries to petro-laborers. The author's forthright responses beautifully reflect the Gulf paradox: the incongruence between the so-called Gulf experience described by the Gulf returnees and the stark reality that the promised land bequeaths to the migrant laborers.

MV: *Many people travel to the Gulf countries for better financial prospects. Some face exploitation in terms of low wages, payments held back for months, physical abuse, confiscation of passports, and so on. But there is hardly any representation of these incidents in creative writing, especially in English. What do you think silences the voices documenting the woes of the Gulf diaspora?*

B: There are several reasons for this. One is lack of education. Those who are facing these types of issues are not well equipped or educated or do not possess any literary background to express and articulate their firsthand experiences. A second reason is fear. Until very recently, censorship of writing was very strict in these countries, so those who wish to write hide many brutal facts due to threats issued by the host country. Another reason is social ostracism. They prefer not to expose the harsh reality of life to people back home. For example, in *Goat Days*, we come across Najeeb's letter to his wife, where it can be noticed that even when he was undergoing great suffering in the desert, he doesn't reveal this reality to her. Collectively, this remains the general attitude of a common Gulf laborer in the petro-diaspora. They prefer to keep the harsh realities of their workplace hidden. They think it might hamper their social status in their home country.

MV: Goat Days *has been translated into Arabic. How has the response been in Saudi Arabia and other Gulf countries? How did the natives of Saudi Arabia react to your novel? Do you think this kind of writing can bring about a positive change in the behavior and outlook towards migrants like Najeeb?*

B: Saudi and UAE immediately banned the novel. It is because of this ban that several natives took to reading the book. Many of them have expressed their solidarity with the fact that such kind of atrocities might have happened in the past. But they disagree and deny that such practices do still exist.

As to the second part of the question, I am really not sure of how much impact it has caused amongst the natives and whether it has transformed their attitude towards the migrants, because most of the people who have read this book in Saudi are well educated, highly placed, sophisticated citizens who more or less refrain from indulging in such kinds of inhumane behavior. If a change has to happen, books like *Goat Days* should reach the people who are semiliterate, and who are appointed in the lower ranks. They are the ones who directly deal with the migrant workers.[1]

MV: *Throughout the novel* Goat Days, *we find Najeeb possessing a very strong faith in God even in the most adverse situations. He never loses hope and surrenders everything to the will of God. As an author, has your spiritual faith in life infiltrated Najeeb's unshakable faith in God?*[2]

B: To move forward in the journey of life, every individual needs an anchor. And this anchor can be anything—an unshakable belief in God, an ideology, faith in oneself. As a novelist, I wanted to put forward this idea of hope to my readers. And perhaps it was the same vision and faith which Najeeb held consistently that led to his miraculous escape from the Masara. Through Najeeb's unwavering faith in Allah, I also wanted to put forth to my readers the significance and power of positive thoughts in the most adverse situations.

MV: *Has this book helped youngsters aspiring to go to Gulf countries now to be a little more conscious about the choice they are making?*

B: A lot of awareness has been generated amongst the youngsters. They have started raising questions like what would be my wage, who is my sponsor, how authentic is my sponsor, [and] so on. This process of seeking clarifications was rare amongst the youngsters emigrating earlier. The government of Kerala too has in a way promoted the book by asking the Gulf aspirants to read the book before going to the Gulf countries.

MV: *Ibrahim Khadiri, the Somalian who helps Najeeb and Hakeem to escape from the Masara, disappears after escorting Najeeb to a safe destination. Many readers have correlated Ibrahim Khadiri with the tiger in the movie* Life of Pi.

Since Ibrahim Khadiri stands as a symbol of hope and inherent goodness in a human being, why do you, as a novelist, choose to make him disappear mysteriously, as it leaves a trail of questions unanswered? Please elaborate.

B: In many interviews and literary functions people have raised the question as to who really Ibrahim Khadiri was. And the answer to this has always been: he is very much a flesh and blood character, one amongst us. He did not do anything extraordinary; he just did what ideally any human being should do when his fellow beings are in distress. The character of Ibrahim Khadiri stands as a symbol of how a human being should help his fellow beings: that is, [in] selfless service without expecting anything in return. Khadiri disappears at the point when he had finished his duty as a guide and a savior, and thereafter he had no further role; hence it was pointless to hold him back in the novel. Paradoxically, in our society, if we look around, we find many individuals who have done meritorious jobs but in return start expecting applauds, grand welcomes, and felicitations, which I feel is a very untoward practice.

MV: *Your works in Malayalam have regularly been published by the best publishing houses in Kerala, like DC Books. But so far only two have been translated into English. Is it because of a lack of good translators or because Indian English publishing houses are not ready for such themes?*

B: Themes of isolation and [the] struggle of migrants in an alien land is not a new theme to readers and to literary studies. In English fiction, there are countless narratives depicting the struggle and the consequent process of acculturation of diasporic communities in host countries. But the experience of [the] labor class in Gulf countries is completely new to readers of English fiction. And it is this unique theme that drew the curiosity of many readers who read and liked the novel very much. It was published by Penguin and they did a commendable job by reaching out to the masses.

MV: *Do you feel that it is the social respectability procured/gained by the migrants after years of hard work that acts as a hindrance to the expression of their Gulf experience? Is there a fear of losing their social standing which prevents them from speaking?*

B: That is absolutely right. They are gaining a social status by the power of money or by the status quo they are maintaining as a Gulf returnee, and they don't want to lose it by revealing the source or hardships endured to acquire the tag.[3]

MV: *Protagonist Najeeb's homecoming hasn't been documented in length. It has been quite terse. Was it done by you intentionally to retain the focus on the harsh realities faced by migrants in the Gulf countries?*

B: I felt that it was not required. The readers can easily gauge [what] the homecoming would be like and what lies in store for a Gulf returnee like Najeeb. Umpteen migrants, despite facing tremendous hardships and hostility in [the] Gulf, have to return [from the] Gulf after a short stint of two years. In real life, too, the plight of Najeeb, whom I unexpectedly met after a few years, is quite similar.[4] He had to erase his past memories and carry on with the business of life. He found it extremely difficult to once again reconcile and accommodate himself to normal circumstances back home in Kerala. Moreover, there have been a lot of stories and films which have already harped on this theme. . . . I felt that it was pointless [elaborating] on Najeeb's homecoming; therefore I left the rest to my reader's imagination.

MV: *A lot of educated and skilled youngsters from states like Kerala go abroad and get hired to do menial work. Although they face a lot of harassment, many still prefer to work in these countries. Has your book* Goat Days *led these youngsters to become aware of the harsh realities of Gulf life? Has it encouraged them to reconsider the choices being made and helped them to opt for employment within India itself?*

B: There was a time back in the sixties and seventies in Kerala when [the] Gulf became a hub for every youngster aspiring to gain social respectability and wealth. If we look at the social reality of Kerala, there was a scarcity of jobs which forced these youngsters to leave their homeland and seek [a] livelihood abroad. There were several men who began with a determined resolution of returning back soon after amassing wealth, but due to pressures back home, they ended [up] extending it to twenty to twenty-five years—that is, almost spending more than half their lifetime working in barren deserts. This trend is slowly declining because we have better-paying jobs available in our own country. Furthermore, if we look at the basic requirement of these oil-rich countries, we would be able to discern the fact that their requirement is basically for unskilled labor, which we no longer have, as most of our youth are well-educated. Therefore, I have a very strong notion that this inclination towards going to the Gulf will further decline in the coming decades.

MV: *Najeeb faces trauma on three major accounts: alienation from home and country, alienation from water, and alienation from language. Which do you think is the most traumatic?*

B: I feel that most painful would be the alienation from home and his most dear ones. Equally traumatic is the experience of the loss of language, which consequently leads to [a] loss of identity. Both of these aspects are very closely interlinked. It is to overcome these losses that, to borrow from Salman Rushdie's term, the migrants end up creating an *imaginary homeland* in the new land. For instance, a person migrating to Burma or [the United Kingdom] assigns the name of his ancestral house to his new home to get a vicarious feel for his homeland. Najeeb fabricates his imaginary homeland by re-creating his village atmosphere. He assigns the goats in Masara the names of people whom he is acquainted with and whom he loves dearly—Sainu, Nabeel, Pocchakari Ramani, EMS, Mohanlal, and so on and so forth. Despite being in isolation, it is this imaginary homeland that gives him the strength and endurance to carry on.

MV: *The perspective and role of Sainu, Najeeb's wife, is very minimal in* Goat Days. *Was this done intentionally? Do you intend to write a novel with a female protagonist voicing her distress about a husband trapped in Masara?*

B: No, at least at this point I am not intending to write one. Even though Sainu occupies very little space in the novel, she is the pillar of support, like any quintessential Malayalee woman who gives endless hope and support to her husband to go abroad. Even at a point when Najeeb is completely disillusioned by the uncertainty of Gulf prospects, it is Sainu's words and support which offer solace and strength to Najeeb to steer [him] out of his delusions.

Sainu is amongst the scores of Malayalee women who played a great role in [the] social engineering of the contemporary Kerala society. It was and still is these women who single-handedly manage the household, take care of the children and aged in the family, and take active participation in channelizing [*sic*] investments and resources to proper places. Thus, they have done an excellent job and on no account [can the] role of women like Sainu be sidelined, as they have an equal contribution in the endeavor to achieve prosperity.

Hence the role of women like Sainu cannot be underestimated, as they play an equally significant role in the strife for prosperity.

MV: *Were you approached by a literary agent? Were you offered a literary advance by the publishers? Do you feel that literary advances hamper the creativity of an artist?*

B: In Malayalam literature, we generally do not have the concept of literary advances. I personally feel that the moment writers agree [to] taking a literary advance, their creativity becomes narrowed down and the writer has to work within the confines of certain boundaries, which I personally don't subscribe to. Creative writing is all about transgressing the preset boundaries to touch the heart of the readers with a message. For example, when a writer is writing from a Gulf country, he/she needs to think about the policies of the country, the reaction or uproar it would generate. This acts as an obstacle to the writing style, as the writer is not able to effectively convey what he/she strongly feels. Again, the time constraints [set] by the publisher can hamper the writer's productivity, as he/she might start scampering to finish the work within the said deadline given by the publisher.

MV: *How did it feel when your novel was banned in Saudi?*

B: It was banned in Saudi and UAE. By the time it was about to be banned, I [had] returned back to India. The most interesting facet about it was that the strongest opposition [to the book] came from Indians, especially the Malayalee business groups situated in Saudi, as they felt it would adversely affect their business prospects.

MV: *Do you feel Najeeb's act of storytelling functions as a catharsis? Do you think in real life, such a process would help in unburdening the trauma faced by the Gulf migrants?*

B: Definitely. Storytelling does possess such a unique quality. For several years, Najeeb was completely cut off from the mainstream world, and it was his insatiable hunger to communicate with people in his own language [that] also unconsciously [became] a conduit to relieve himself from all the accumulated impressions of pain and trauma he was carrying.

Additionally, Najeeb was facing the risk of losing his own language as a result of disuse. I have come across several newspaper accounts where migrants have related their misfortune of forgetting their own native language after several years of residence in the host country. The loss of language is almost similar to erasure and fragmentation of one's own identity, which becomes quite painful. Furthermore, many of them are not able to openly articulate the inhumanities they have faced even to their immediate family members because of the fear of inflicting sorrow on them.

MV: *Has your training as an engineer aided you in the process of creative writing?*
B: An engineer can meticulously design and plan things. He has the foresight to look into the repercussions of placing a particular incident in a particular sequence. Being a mechanical engineer has tremendously helped me to organize my thoughts and in planning the structure of the plot.

MV: *Every writer has a particular schedule and peculiar fancies, like a particular chair, pen, locale, and so on. What is your writing schedule like, and do you have any such writing fancies?*
B: The most comfortable place for me to write is my study, where I feel extremely relaxed and comfortable. I generally write after 10 p.m., which I feel is very convenient, as there are [fewer] distractions.

Conclusion

For many youngsters, the Gulf has represented an ultimate hub of high employment, prosperity, upward mobility, and cosmopolitanism. Since the eighth century, Kerala has fostered strong trade relations with the Arabs who come to the Malabar Coast for spices and pepper. Even today, Kerala is a major producer of spices such as cardamom, cinnamon, clove, turmeric, nutmeg, and vanilla that form the cash crops of the state. This bilateral Indo-Arab trade fostered years ago made a deep historical mark and penetrating imprint on the Malayalee imagination. The 1960s and 1970s in particular saw an exodus of young men leaving the state to explore their fortunes in the Middle East. The euphoric bubble of the postindependence sociopolitical scene (around the 1950s), which frothed with enthusiasm, soon collapsed in Kerala when the youngsters witnessed a wave of poverty, hunger, and caste-based oppression. The state couldn't offer much in terms of employment, and with the discovery of the Middle East as the *land of black gold*, more and more young people aspired to emigrate to West Asian countries, pawning whatever little ancestral property or gold they possessed to get there. By the 1980s, the Gulf remittances contributed an average of 30 percent to rural household incomes. A lot of illegal migration took place during the 1970s and 1980s, with dubious agencies creating fake passports and arranging visiting visas for these aspirants. The word "Gulf" for a commoner in Kerala meant Saudi Arabia, Dubai, Kuwait, Qatar, Oman, or Bahrain. Interestingly, the earliest migrants who emigrated from the lush green topography of Kerala faced a crude culture shock, and it took them months to grapple with the scarcity of water and vegetation. Most of these initial migrants were semiliterate; they held certificates from ITI polytechnics or had undergone short-term certificate courses for positions as mechanics, welders, or radiographers. And perhaps it is this lack of education that hindered their ability to narrativize their struggle in literary fiction or through creative mediums.

The remittance economy from the Gulf countries is one of the factors that helped Kerala become the first state in India to achieve extremely high literacy rates. The remittance sent back home was ostentatiously displayed to help them climb the social ladder. The migrants mainly invested in building palatial mansions, owning flashy cars, and educating their children in convent or boarding schools.[5] They encouraged their children to take up professional courses like B. Tech or MBBS, which acted as catalysts to further enhance their social status. From the 1980s onward, Malayalam films such as *Vilkkanundu Swapnangal* (1980), *Visa* (1983), and *Akkare Ninnoru Maran* (1985) contributed significantly to perpetuating the Gulf allure. The hero returned from the Gulf was accorded an iconic status worthy of emulation and worship. They were represented as possessing glossy symbols of power including a car, dark glasses, a color TV, a telephone, a Panasonic tape recorder, suitcases full of perfume, and imported liquor for their friends and relatives. These men became the most sought-after bridegrooms, granting them the opportunity to choose from many beautiful and educated brides.

Another interesting reason that many of these youngsters preferred emigrating to Gulf countries was that it was easier than emigrating to the European countries, where the selection criteria for employment was more stringent. These emigrants tried to re-create a feeling of home in the host country in myriad ways, like playing songs from Malayalam movies, celebrating festivals, reading Malayalam magazines, and having food in restaurants that specifically catered to the taste buds of Malayalee Gulf emigrants. In fact, they displayed an exemplary unity in their struggle to keep their woes hidden from their loved ones and to retain the facade of prosperity in their homeland. Furthermore, many of these workers collectively pooled money and created small chit funds to give to those who were going home on their annual visits. In fact, Professor Irudaya Rajan of the Centre for Development Studies puts forth the claim that "without the Gulf migration, high levels of unemployment and poverty would have made Kerala a hotbed of terrorism, communalism and social tensions." He adds,

> The impact of remittances on Kerala is manifested in household consumption, saving and investment, the quality of houses, and the possession of modern consumer durables. Remittances also play a major role in enhancing the quality of life and contribute to a high human development index for Kerala in terms of education and health, along with the reduction of poverty and unemployment. The Gulf, which is home to a large majority of emigrants from Kerala, has figured prominently in this equation.[6]

Thus, emigration had allowed for a quantum improvement in the standard of living.

In terms of gender relations, many communities in Kerala practiced the unique matrilineal system whereby the property rights of a household were passed on to the female descendants. But this system of kinship disintegrated in the early twentieth century, leading to high levels of social and political conflict. By 1981, female literacy in Kerala was well over twice and infant mortality less than a third of the rates for India as a whole. Although the state was successful in offering good education to women, it couldn't offer much in terms of employment within the state. Many women got married to young men working in the Gulf. They couldn't accompany their husbands to the Gulf, as many of the men lived in labor camps or simply couldn't afford to take their wives with them. Women like Sainu (Najeeb's wife) in *Goat Days* single-handedly took care of the house, managed property and remittance money, and also cared for the children and elders in the family.

Benyamin's novel *Goat Days* has contributed immensely to engaging a broader audience and making them realize the reality behind the glitz and glamour of the Gulf. The protagonist Najeeb employs storytelling as a medium for multiple purposes: as an agent of catharsis, as a means to regain his lost identity, as a means to burst the Gulf lure, and as a means to foster new ties with prisoners in Bath. The Brechtian technique of clinical representation, shorn of all romanticism, paves the way for reflection on the intensity of identity crisis Najeeb would have felt in Masara. Although writer Benyamin has shown the state as a party to the nefarious practice of frothing this Gulf dream, the novel was well received by the Kerala government, and in fact it was suggested by the former minister of Overseas Indian Affairs, Mr. Vayalar Ravi, that it is a must-read for all the young people who were aspiring to go abroad. In fact, the novel has succeeded in capturing the attention of a worldwide audience; it has been thoroughly read and discussed in book festivals in especially Karachi (Pakistan) and Dhaka (Bangladesh), since there are a lot of laborers emigrating from these countries to the Gulf. So far, one hundred editions have been printed, making it the one of the most popular works in the Malayalam language.

In the postcolonial era, works of writers such as Benyamin help us to rethink as well as challenge the notion put forth by Western academia that most of the diasporic experiences can be conveniently interpreted within the paradigms of the Western diasporic epistemology. Benyamin's *Goat Days* not only engages with exposing the myth and falsification of the Gulf dream vehemently put forth by media but also points out how the state can progress from its passive stand and play a crucial role in generating awareness of the woes faced by Gulf migrants. He firmly believes that his writings could propel youngsters to rethink the options and choices they make in life.

It was indeed a wonderful experience visiting his ancestral house and talking to him on various issues, but above all, it's his commitment to writing, his knack for engaging in spirited discussions, and the hope he nurtures for a brighter future with more favorable foreign policies for Gulf employees, that enthrall his readers and interviewers alike.

Notes

The interview with Benyamin took place on June 6, 2016 at his residence in Pandalam, Kerala, India.

1. The managers who are appointed to oversee the work done by migrant laborers are locals, and they possess minimal educational qualifications. Here, the author refers to the men who occupy these supervisory ranks.

2. The question regarding faith was raised in the interview; as in most creative forms (especially novels and plays), godly intervention / lack of intervention and faith / lack of faith have been represented as lending or denying assistance in mitigating a crisis. For migrants, it is strength of faith that plays a significant role, along with others, in determining the endurance

of crisis in a hostile environment, as in *Goat Days.*

3. In many cases, although these workers were hired for specific jobs, they ended up working as cleaners, security guards, and so on. Due to the social stigma associated with these jobs, they concealed them from their families, who basked in the illusory glory of their newfound wealth.

4. The plot of the story is based on the real life of a Gulf returnee named Najeeb whom the author encountered accidentally; Najeeb's experiences triggered the flow of events in the novel.

5. In their book *Kerala's Gulf Connection, 1998–2011: Economic and Social Impact of Migration*, noted social scientists K. C. Zachariah and S. Irudaya Rajan remark that a major impact of the remittance was on the quality of houses in Kerala: "It is a common sight to see lines of palatial houses even in remote rural areas; an indication that many individuals from that area had once been migrants" (Hyderabad: Orient Blackswan, 2012), 83.

6. For more details, see http://www.mei.edu/content/remittances-kerala-impact-economy.

15.

Testimonies from the Permian Basin

Kristen Figgins,
Rebecca Babcock,
and Sheena Stief

In many ways, American culture is built on the back of the boomtown, which is itself symbolic of the American dream. Every year, semimigrant workers travel across the country to boomtowns—settlements (both large and small) where economic advancement is promised. As oil production ebbs and flows and prices rise and fall, urban centers such as Odessa and Midland, Texas—part of a geographic region known as the Permian Basin, arguably the largest oil-producing region in the United States—swing from industrial catharsis with rapid expansion and population gains to economic devastation, leaving buildings abandoned, businesses shuttered, and huge numbers of the population without work. As discussions of energy production increasingly shift to concerns about climate change, this discussion becomes more a part of the cultural landscape of the Permian Basin and of the United States in general. The Permian Basin community grapples with thorny questions about whether the experiences of people in boomtowns are totally unique. Many members of the community have normalized their experiences and are surprised to learn that the relationships they foster with employers, with neighbors, and even with their environment are atypical compared to those in other urban centers.

In this chapter, we will use a collection of sustained memoirs from the Permian Basin community as a starting point for exploring the impact of energy production on the boom-bust ecosystem. We have already conducted research into sustained narratives from members of the Permian Basin community as part of a Humanities Initiatives at Hispanic-Serving Institutions grant from the National Endowment for the Humanities (NEH). Our analysis will pull together the discrete narratives about the human and environmental encounters in this community and attempt to unify them through their perspectives on the boom-bust ecosystem.

About the Boom or Bust Project

The Boom or Bust project began as a grant initiative to serve the Permian Basin community; it was conceived by Rebecca Babcock, Kristen Figgins, and Jason Lagapa, all from the University of Texas Permian Basin (known locally as UTPB). In the fall of 2016, Rebecca asked for volunteers to try for a Humanities Initiatives at Hispanic-Serving Institutions grant, and Jason and Kristen volunteered. At the time they wrote the grant, the area was on the tail end of a bust, so the petro-economy and its incredible impact on our community was particularly highlighted as we began our talks. In the two years that the project ran, the fortunes of the Permian Basin shifted, and it began experiencing one of its biggest booms, which shows just how quickly the tables can turn in boomtowns.

In our early talks, we discussed wanting to do something that would be impactful to our region. We were going through a bust—the oil field company Kristen's husband had been working for had just gone under—and our talks kept coming around to the energy industry in the Permian Basin, how atypical it is, and how so few people seem to talk about it here in a really focused way. Boom or Bust evolved out of those talks and became a plan to get people in our community to engage with the energy industry and really talk about it. In addition to creative-nonfiction writing workshops, from which many of our testimonies came, we also developed book clubs and a speaker series. One of our goals was to be ready for the next boom, to preserve the wisdom of people who had been through the cycle before and could share their perspectives for the next wave of the cycle.

Eventually, we received news that the Boom or Bust project had been funded, and we began to enact our plan to foster discussion about the field of energy among residents of the Permian Basin. At the end of 2017 academic year, Kristen left the project to pursue her PhD in Arkansas, and Sheena Stief was brought on as project coordinator. Sheena was a natural choice, because as someone who was raised in the Permian Basin she has always been interested in what people had to say about energy in the area, especially oil. Like Kristen, Sheena had family who worked directly in the oil field and on the periphery.

As the project developed, the coordinators were surprised to learn that their narratives were emerging as part of the project as well. Rebecca came to the Permian Basin in 2005 for her job—the same job she has now, except that she has since gotten two promotions. Kristen was born in Odessa (although at the time of her birth, her family lived in Hobbs, New Mexico, also a boomtown), and she has lived in Odessa all her life, except for two years when she lived in another, very different oil field town (Lafayette, Louisiana) while pursuing her master's degree. Sheena was born in Abilene, Texas, and moved to Big Spring, Texas, when she was two. She grew up in Big Spring and moved to Midland when she was twenty. She's been in Midland for the last thirteen years. Although their backgrounds were different, each of the coordinators had been in the area long enough to experience multiple booms, multiple busts, and how they impacted the area.

The importance of the project was highlighted in the fact that few people in boomtowns share their experiences in a focused way. While talk about the economics and environmental impact of the energy industry is plentiful, both in popular media and in academia, few studies focus on how the people of boomtowns are affected. In addition, communities do little work to allow people to share the human side of the story of oil and energy production. Each area in which oil is found around the United States has its own set of people. It's important to capture the many types of people and their voices—flaws and all. These stories quickly became one of the biggest successes of the Boom or Bust project. As we learned, when you live in a boomtown too long, people tend to normalize things; they complain about the uptick in traffic during booms and the lack of jobs during busts, but they don't always think about the other ways life in a boomtown is really different. One of the really obvious effects of this inattention is that during the booms you'll see people (as people in our testimonies noticed) spending money as though the boom will never end; it's one of these missteps that occurs because so much of what makes boomtown life work goes unspoken by the people who have lived there long-term. This project is important because people need to hear one another's voices on this subject instead of what they usually hear: the voices of journalists or oil field executives.

In fact, although the project was run through our capacities within the University of Texas Permian Basin, the idea of hearing one another's voices was important not just for students and professors but for everyone in our community. People from all demographics were invited and attended our events, participating in the writing workshops, book clubs, or speaker series. Each of us found ourselves enjoying particular moments. Rebecca's favorite moment was going to the petroleum museum with Stephanie LeMenager, a leading expert in Environmental Humanities, whose book *Living Oil* was one of our favorite book club readings. For Kristen, the project was a general wake-up call that she wanted to study ecocriticism: it was the first time she realized how energy production dominates our lives; working on the grant and talking to people as part of the program was a huge part of that realization. Sheena sums up the experience well when she says that she really enjoyed working with and listening to the many types of narratives. People are so interesting, and it's been great experiencing what they have to say and how they say and write it.

At this point, the grant is complete, but our role in recording the testimonies of the Permian Basin is ongoing. We have published a book, *Boom or Bust: Narrative, Life, and Culture from the West Texas Oil Patch*, with Oklahoma University Press that includes the narratives of those people who participated in our project. We hope that this will have many benefits, including that its readers will learn from it and enjoy it. Those whose narratives are included in the book will feel pride and a sense of accomplishment in sharing their stories, but we're also excited that these voices can be heard louder and farther away. Learning what local people have to say about local issues will benefit not only local individuals but also those in academia and those involved in energy production around the United States.

Development and Collection of Testimonies

Part of the NEH project was a creative-nonfiction writing workshop, designed to let the Permian Basin community share their experiences in their own words. These workshops were led by Dr. Jason Lagapa

> and focused on both work and living experiences during economic downturns and upticks in the region. Workshops began with a writing prompt, and participants would write for fifteen to twenty minutes. Afterwards, participants would discuss their writing, receiving feedback from the rest of the group. I would also offer my own impression of the written pieces, giving encouragement and advice for revision where needed.[1]

The workshop led to many partially developed testimonies, and participants were encouraged to send those to us for archiving. Some of the testimonies developed in the project were read aloud by the authors and broadcast on Marfa Public Radio, where they were archived. At some point in the project, however, we recognized that while people were very interested in sharing their stories, not everyone was interested in publication.

At this time, we began to solicit testimonies from participants in the project and from the broader community for inclusion in a larger book project. We conceived a collection, titled *Boom or Bust: Narrative, Life, and Culture from the West Texas Oil Patch*, of the narratives along with scholarly articles on the topics of the boomtown and of boom-bust cycles as they relate to energy production in the Permian Basin. As of this time, we have collected around twenty sustained written narratives. The people who submitted narratives came from diverse backgrounds. Some of the writers were young people—high-school and college students; others were adults reflecting on the ups and downs of their careers in the oil industry. Still others were employees in workplaces such as restaurants and schools.

Generally speaking, few parameters were given regarding the narratives. At no time during the solicitation of these pieces did we attempt to regulate the length or content of the testimonies using any specific metrics. If we felt that participants were scratching the surface of something that deserved further exploration in their narrative, we encouraged them to write more, but we did not require such expansion for inclusion. Despite this, some definite themes began to emerge from the project.

Testimonies and Their Impact

Many of the narratives featured discussions about traffic. From living and working in the area, all three of us have noticed that traffic increases over time when the boom is on. Sheena Stief's narrative is most directly about traffic, as she discusses an incident she had running over a box in the road that contained an air filter and her car subsequently catching fire. In this narrative, the oil field is both the hero and the villain. Negligent (or unaware?) oil field workers allowed the oil filter in the box to fall to the road and did not retrieve it; or perhaps they did not know it fell. Running over the box, though, caused Sheena's vehicle to catch fire. Oil field workers stopped to help her, to extinguish the fire, and to rescue her laptop from the burning van. She writes, "The oil industry caused the incident, but stepped in when I needed it most."[2]

Alex Rathbun grew up in a rural area and noticed that when the boom was on, "it got noisier every day. Trucks would drive by more frequently; the truck stop down the road became more crowded." She also notes that her parents got annoyed with "those guys and their trucks."[3] Contributors provided both positive and negative experiences with the influx of traffic and the types of vehicles on the road. Rathbun detailed not only her experiences but also those of her parents. Although Sheena and Alex both chose the topic of traffic, their focuses were very different. The ebb and flow of the narratives is much like the ebb and flow of the Permian Basin oil industry: a common theme but not a common experience.

The focus on traffic, roads, and cars is perhaps unsurprising. Just as the petroleum industry focuses on producing rubber, gasoline, and asphalt, so too does the industry rely on these things. One of the first signs that a boom is oncoming in the Permian Basin is the sudden influx of vehicles, both those that service the industry and those that are by-products of the influx of workers and families supporting that same industry. Since the infrastructure of the Permian Basin is not built to accommodate the inundation of transportation, it's felt very viscerally by the residents of the area. The physical reality of the traffic is not the primary concern in most of our testimonies, however. Equal attention is given to the quality of the driving. As Daniella Garcia puts it in her narrative, "The traffic gets nuts[;] it's scary to be on the road because everyone is driving like [a] maniac, and yes that includes both you and me, but I think that happens regardless of booming or busting because the roads are just that bad."[4] During the boom of 2014, KXWT West Texas Public Radio published a report that car fatalities had risen by 13 percent from 2012 to 2013, which reporter Tom Michael saw as directly tied to the boom. Michael's interviews with Midland County Sheriff Gary Painter seem to support that: "Painter says these newcomers [oil field truck drivers] can stand to make anywhere from $72,000 to $100,000 a year, but just because they get the job doesn't always mean they're qualified for it."[5] As Painter notes, a career in the oil industry is unlike many other industries in that everyone, even if they work in support fields, is directly affected by a boom or bust.

A number of the narratives either focused on or mentioned careers in the oil field or jobs outside of the oil field. Working in a town that moves with booms and busts becomes tricky, according to our contributors, whether or not one is working directly in the oil field. For example, Katie Groneman discussed the process she went through to find work as she transitioned from college to the workforce in the Permian Basin. Groneman graduated from the University of Texas Permian Basin and sought out work after graduation. She ended up working for a retail store until she could gain more permanent employment as a high-school teacher.

Whereas Groneman discussed her transition from college student to full-time employee, Dave Stief discussed how he began working in the Permian Basin oil field and how variable that type of employment is. Stief described his experience as "a very difficult and tremendously unpredictable professional and personal journey."[6] Stief further described his work experiences this way:

> We worked 24/7 when on call as a service company because the cost and ability to evaluate oil and gas wells dictated schedules. We had a short window to help determine a well's commercial potential. I generally had a three-person crew. Two people to operate the truck and I had a car. Once we were headed to a well it was our job to evaluate until the job was complete, no matter how long it took. In my area at the time you might spend 5 hours or 25 hours at a wellsite. Rain or shine, night or day, hot or cold and on any day of the week we were needed to help customers evaluate their wells.[7]

Stief's experiences were echoed in the book clubs for the project, especially as we read *Oil!* Upton Sinclair's novel on socialism and labor conditions in the oil field was a difficult read for many of our participants, who were generally surprised to find that the harsh conditions of the industry haven't changed much since Sinclair published his novel in 1927. Like Stief, the crews who operate the rigs in the novel worked long hours in rough conditions. Participants in our book club also noted that the poor safety regulations of *Oil!* are echoed in today's environment. Similarly, Stief recounts the difficulty in getting food and sleep during his time in the oil field:

> We took food and drinks to help keep us going as in a lot of cases we weren't near a restaurant or didn't have time to go get what we needed to fully enjoy the experience once the work at the well began. There was a sleeper berth in the truck for a single person while on location and I had the back seat of a car. The truck crew would try to alternate sleep periods, if possible, but the engineer was up from beginning to end once the logging started. Once we finished on any given well it was our job

> to get everyone home safely and prepare the truck for the next job or directly proceed to help customers on the next well. Too many times my crew and I were less than overjoyed to hear that we had another well waiting on us. But that is what we signed up for and away we went.[8]

Despite the challenges, Stief went on to mention that he has enjoyed his job overall. Careers and people's experiences with their careers fluctuate, whether they work directly in the oil industry or in the periphery.

Meanwhile, Bryce Heckman described company layoffs that took place in 2016. Heckman describes the anxiety and feelings of loss and guilt as he and his coworkers struggled to survive in the company: "By noon the layoffs were over. Back to work, the e-mail said. Rumors circulated. Last I heard it had been a company-wide twenty-percent reduction. Regulatory was now a group of twelve."[9] Heckman's experiences are incredibly common, especially in a bust. In 2016, the *Texas Tribune* remarked that "thousands have been laid off. Tax collections are plummeting. Many are on the brink of homelessness. Rows of drilling rigs and white company trucks sit idle—there's no telling for how long."[10] This is the opening of a relatively optimistic article about the resiliency of the West Texas oil industry.

Another contributor, Dennis Robbins, went into detail about his experiences working in the Permian Basin oil field. Robbins experienced what it was like to be laid off in the big oil bust of the 1980s, an event that is still discussed in the Permian Basin.

> Hundreds of geologists, geophysicists, landmen and engineers lost their jobs when OPEC flooded the market with oil. It's easy to really get discouraged after getting let go from a job, and to keep things in perspective, I looked at the folks I had been laid off with from Gulf, and realized it really wasn't personal. These individuals were some of the best in the industry and the companies were making decisions based on economics. We were working in an industry that had to adjust to low oil prices. Company loyalty was definitely undermined, as the 30-year career folks were a thing of the past.[11]

Robbins goes on to explain the difficulty of finding a new job during the bust, when hundreds of other individuals are trying to snatch up those same jobs. It's an experience that still rings true today; in that same 2016 *Texas Tribune* article, Leslie Kinney shared how she and her husband had been living in a storage unit; her husband couldn't find work, even in the fast-food industry: "He's tried to find work at local fast-food restaurants, but has been told he's overqualified," wrote *Texas Tribune* reporter Kiah Collier.[12]

But as Robbins shares his experiences in his own words, it becomes immediately apparent that things have also changed in the nearly thirty years between the two busts. According to Robbins, the common thing to do at the time was to "go door-to-door and visit with everyone I could,"[13] even though companies today encourage interested applicants to submit resumes via email or online application:

> I followed his advice and began by going downtown . . . to a building marquee, and [writing] down every company that sounded like it had something to do with oil. I would start on the top floor and drop off a resume to every company who would take one. I would try to visit two buildings a day, starting at 9 a.m., ending with the first building before 11. (Because everyone is out to lunch after 11.) Then I'd repeat the process in the second building after lunch. . . . I guess I visited approximately 70 different companies and went through almost every building in downtown Midland.[14]

In our roles at the University of Texas Permian Basin, it was not uncommon to see ex–oil field workers returning to school in order to diversify or enhance their resumes. Like Robbins, many of our students at the university reported feeling uncertain about their prospects in the oil field. Others were certain that the boom would return (as it did). Our contributors highlight that a core feature of the industry, which must be recognized by future hopeful generations, is that it is inherently unstable.

Some of our contributors were interested in transition periods. For example, Berry Simpson was offered a promotion and transfer that did not go through during the bust of 1986, when he saw friends losing their jobs.[15] Maggie Luhan wrote of her mother getting laid off and being okay, because her family had other resources, but also of hearing stories in school about families losing their cars or having to move for other jobs.[16] While discussing her career, Edith Vandervoort spoke of the importance (as did Corni Ortega, a high-school student)[17] of saving money and spending it wisely between the ups and downs of the Permian Basin's oil industry. Vandervoort had several places of employment throughout her working career, and she details why money should be spent wisely during the boom: "Despite the added income from our leased properties, I had to withdraw even more money from my retirement account with a ten percent penalty. I was aware of that, but I did not realize that the money from my IRA is counted as income and would put me in a higher income tax bracket. Now I owe the IRS quite a bit of money, but because we are in a bust cycle again, things are getting better, slowly."[18] Vandervoort worked directly in the oil field through a privately owned business and, later, indirectly by teaching, which she describes as being an occupation that was affected by the oil industry as classroom sizes increased. Her narrative moves from her disappointment in the bust cycle to the larger disappointments she felt in the government and especially regarding oil's cumulative effect on the environment.

Almost all of the collected testimonies mention working either directly or indirectly in the Permian Basin's oil field. Everyone's careers in the Permian Basin are impacted in some way by the booms and busts. After analyzing the narratives, we realized that everyone walked away learning some kind of lesson from their experiences. Some learned the importance of saving money because careers don't always last, while others learned the importance of working together and persevering despite a boom or a bust.

For example, Chris Bartlett writes about his dad working long hours and consequently his really only getting to spend time with his dad in oil field shops: "For better or for worse I grew up in those shops my dad worked in when I was a kid. . . . I would spend my Saturdays in the shop where he worked since I didn't see him during the week." He also wrote about "[losing] touch with many friends throughout the years as their families moved from the desert to greener pastures during times of busts."[19] The shop hands, though, acted like they were his "temporary siblings." Bartlett portrays the oil field shop as a home where real and surrogate family relationships were forged. Bartlett's experiences were not unique in our testimonies. Corni Ortega writes about how her "firstborn moments were spent with [her] father away at location."[20] She wrote about how her dad "worked all week, sometimes not sleeping for four whole days. When [she] was born, though, he took off three months to be near his firstborn daughter." She concludes with, "I thank the oil field every day for the opportunities it has brought my family."[21] Jessica Terrell's testimony was a tribute to her grandfather, showing again a pattern of our contributors highlighting their deep respect for their families as they reflect on their hard work and achievements in the oil industry.[22]

Chris Bartlett categorizes the desire to work in the oil field as people "working to make a better life for themselves and usually their families as well."[23] What many of our contributors define as a better life, however, includes things that can be hard to get, no matter what part of the cycle they're experiencing. Basic material needs are sometimes difficult to come by during both booms and busts in the Permian Basin, something that is shocking for many people new to the area. The most basic needs, such as clothing, food, and shelter, are scarce or prohibitively expensive. During the Permian Basin's booms and busts, contributors mentioned both the good and the bad of accumulating the necessities for survival. Accumulation of such items was not the only topic of discussion when it came to basic needs; how those items were used or valued was also of importance. For example, Dezmon Goobi wrote about the incredibly long lines at restaurants during the boom, which made eating out difficult.[24] Even shelter was hard to come by during the boom. Katie Groneman talked about the hardships of trying to find housing in the Permian Basin. Groneman graduated from the University of Texas Permian Basin, and once she obtained her degree, she had to leave the dorms. Groneman found it difficult to secure somewhere to live during the boom.[25]

Not only was it hard for her to find a dwelling, but once she did find one, the price was almost too much for her to afford right out of college. Groneman described her experience:

> 2013 was a good year for the oil industry. For a recent college graduate, it was a nightmare. All the oil companies rented out any available apartments. Demand encouraged new construction. The new apartments sat pristine on the barely developed 191. They mocked me though with their monthly asking rent of $1,500! Who could possibly afford that?! Oh wait, oilfield workers could. Unfortunately, I, like many other Odessa residents, did not work in the oilfield and consequently did not make oilfield money. I went all over town trying to find an apartment that was in budget and available. After the fifteenth apartment and no luck, I broke down. Being an adult aspiring to be a teacher in the oil industry was tough. Luckily, I had a friend who cosigned on an apartment, but he had to prove he made three times the rent. I moved into my 500 square [foot] apartment that cost me $850 a month, but I now had a home.[26]

Berry Simpson, a contributor who delivered his narrative over West Texas Public Radio, described the housing market as crashing during a downturn and a friend of his getting a job at Albertson's as a meat slicer, allowing him to keep his house. In instances such as these, finding housing wasn't the difficult part—it was continuing to pay for the shelter you did manage to find.[27] As in the *Texas Tribune* story, it was not unusual to hear of people living in tents or storage units or couch surfing during a boom. Alex Rathbun wrote about her parents splitting off their land to sell lots to "the crazy people who would want to live out here in this shithole."[28] She wrote about moving into town but then going back to see her old place: "Our area used to be such a beautiful and peaceful place, but every time I drive by my old house and neighborhood it gets trashier and uglier" from the oil field activity.[29] Despite fluctuations, basic necessities are just that: necessities. They must be acquired and maintained. The boom-bust cycle does not allow that even something as simple as housing is guaranteed.

Kay Kolb, meanwhile, in her heartbreaking testimony about reporting on the double murders of two women during the tail end of the boom in the 1980s, remembers that the influx of workers into the area brought different economic expectations. Like today, housing became short during the boom. Kolb remembers:

> During her spring semester sociology class, one woman asked her professor about the trucks. "Are these people all coming from so far up north they have to have those big trucks to drive through the ice and snow?" His reply was not what would be expected. "The trucks are their status symbol. They can't get a nice house to show that they are good neighbors, so the trucks become their home away from home." It was something we hadn't considered.[30]

The testimonies collected as part of the Boom or Bust project revealed a remarkable consistency as our contributors recalled booms and busts, both in recent and distant memory. One would hope that, as newcomers to the area seek out their fortunes, they would find themselves reading about the hardships that may await them, not just the possibilities for success. These testimonies are bracing warnings, even as they encourage resiliency, optimism, and faith that things will always get better.

Throughout the narratives, there was an enduring sense of place. Chris Bartlett wrote that he hated growing up in Odessa and that "hating where you have lived most of your life is a fairly natural experience," but later he reflects that "I stay in Odessa most likely from some sense of civic duty to make it a better place."[31] Almost none of our contributors admitted to enjoying the Permian Basin, although Daniella Garcia allowed that "the sunsets are beautiful.

Bright blues and burnt oranges set all in the background with neon pink and purple ribbon-like stripes piercing through the big sky that never ends. There ain't nothing like West Texas Sky."[32] Despite persistently negative descriptions of the land, the people, the economy, and the cycles themselves, the people here have a definite sense of community. As part of this project, we have come to believe that one thing that unites this community is that they have experienced something truly extraordinary. Boomtowns might be the name of the game across America, but the energy industry brings challenges and opportunities almost unparalleled by any other industry. The highs and lows that the people of the Permian Basin experience are, we think, very much worth recording, revisiting, and (perhaps especially) studying. Although academia may always be more drawn to the flashy and timely aspects of the energy industry, like its economics, we find the slow accumulation of data from its community a worthy endeavor and one that might prove useful to future generations of this unique boom community.

Notes

1. Jason Lagapa, email message to author, 2018.
2. Sheena Stief, "The Dance" (unpublished manuscript, 2018), Microsoft Word file.
3. Alex Rathbun was featured on *West Texas Talk*, hosted by Diana Nguyen. "Personal Narratives from UTPB's Boom or Bust Writing Workshop," *West Texas Talk*, NPR, Marfa, TX: KXWT, August 23, 2017.
4. Daniella Garcia, "432 Township" (unpublished manuscript, 2018), Microsoft Word file.
5. Tom Michael, "Traffic Fatalities Still on the Rise in West Texas as Drilling Surges," *StateImpact Texas*, NPR, Marfa, TX: KXWT, April 22, 2014.
6. Dave Stief, "Dave's Oilfield Narrative" (unpublished manuscript, 2018), Microsoft Word file.
7. Ibid.
8. Ibid.
9. Bryce Heckman, "Two Co-workers" (unpublished manuscript, 2018), Microsoft Word file.
10. Kiah Collier, "Despite Oil Bust, Midland Is Still Bustling," *Texas Tribune*, June 1, 2016, https://www.texastribune.org/2016/06/01/midland-leaders-confident-bust-has-bottomed-out/.
11. Dennis Robbins, "Oil Industry Ups and Downs" (unpublished manuscript, 2018), Microsoft Word file.
12. Collier, "Despite Oil Bust."
13. Robbins, "Oil Industry Ups and Downs."
14. Ibid.
15. Berry Simpson was highlighted on *West Texas Talk*, hosted by Diana Nguyen. "Personal Narratives from UTPB's Boom or Bust Writing Workshop," *West Texas Talk*, NPR, Marfa, TX: KXWT, August 23, 2017.
16. Maggie Luhan and Corni Ortega were both featured on an episode of *West Texas Talk*, hosted by Diana Nguyen. "UTPB's Boom or Bust Project Collaborates with Permian High School," *West Texas Talk*, NPR, Marfa, TX: KXWT, November 30, 2017.

17. Nguyen, "UTPB's Boom or Bust Project."
18. Edith Vandervoort, "The IRS" (unpublished manuscript, 2017), Microsoft Word file.
19. Chris Bartlett was featured on *West Texas Talk*, hosted by Diana Nguyen. "Personal Narratives from UTPB's Boom or Bust Writing Workshop," *West Texas Talk*, NPR, Marfa, TX: KXWT, August 23, 2017.
20. Nguyen, "UTPB's Boom or Bust Project."
21. Ibid.
22. Jessica Terrell, "Earnest" (unpublished manuscript, 2018), Microsoft Word file.
23. Bartlett, in Nguyen, "Personal Narratives."
24. Dezmon Goobi, "UTPB's Boom or Bust Project Collaborates with Permian High School," *West Texas Talk*, NPR, Marfa, TX: KXWT, November 30, 2017.
25. Katie Groneman, "Boom or Bust" (unpublished manuscript, 2018).
26. Ibid.
27. Simpson, in Nguyen, "Personal Narratives."
28. Rathbun, in Nguyen, "Personal Narratives."
29. Ibid.
30. Kay Kolb, "My Heart Remembers" (unpublished manuscript, 2018).
31. Bartlett, in Nguyen, "Personal Narratives."
32. Garcia, "432 Township."

Afterword

Imre Szeman

It has become impossible today to avoid the inclusion of energy in our discussions of, well, everything. Most pressingly, of course, energy has become linked to the politics of our environmental futures. The continued use of fossil fuels (at an ever-increasing rate: in 2017, global use totaled 98 million barrels of oil per day, or almost 36 billion barrels per year)[1] constitutes one of the major barriers to mitigating or addressing climate change. The burning of fossil fuels leads to the emission of carbon dioxide, which leads in turn to the warming of the planet, at a speed that is proving to be faster (and the next time we look, faster again) than we might have thought or hoped.[2] The equation oil plus heat equals CO_2 is easy to understand. Much harder to imagine is how to act on climate change given the largely unacknowledged commitments we have made to fossil fuels in our built infrastructure, in the mechanics of our everyday lives, and even in those dreams for freedom and autonomy that animate individuals and communities for whom neither has ever been easy to access and sustain.

The struggle over the fuels used and the scale at which they are used as well as the fight over what might come next has only just begun. As I write this at the beginning of 2019, the *gilets jaunes* have once again occupied the streets of Paris; their protests have been mimicked near Edmonton, Alberta, fueled by "extra rightwing populism" on the part of those assembled there in *support* of Canada's flagging oil industry.[3] Fuel has a different significance for Parisians from the *banlieues* than for exurban Edmontonians, and it is different again for the Wet'suwet'en people who were arrested in January 2019 for protesting the construction of a pipeline across their unceded territory.[4] These various protests highlight how energy is connected to class, identity, community, work, and politics in complex and distinct ways, as well as how there are stresses and strains on being and belonging as we move deeper into an era that has been named a "petroculture."[5]

One of the sustained critiques of existing discourses on environmental futures—narratives intended to produce generative outcomes in relation to global warming—has been their tendency to flatten the globe, eliminating in the process essential differences in relations to the environment and the impact of climate change (not to mention recognition of the uneven causes of global warming). This flattening has been one of the key outcomes of the numbers game that has guided discussions of the environment to date, with figures such as 400 ppm, +2°C, or +1.5°C commanding our attention. Such universalizing of the experience of global warming has also threatened to be the outcome of the terms and concepts—the Anthropocene most notably among them—that we have hoped would refashion our sensibilities, only to assert in the end a bad eco-cosmopolitanism that fails to truly attend to what is experienced on the ground. Critics have alertly drawn our attention to the very real differences in climate change as it is experienced around the world. So too have communities who have felt disregarded by globalizing elites who can see the problem only from a narrow perspective—a vantage point that might be said to have helped generate the problem of climate change in the first place. For most inhabitants of the Global North, climate change has yet to threaten lives and livelihoods enough to propel individuals to become climate refugees; for many others across the world, however, a small increase in regional temperatures has meant that it is no longer possible to grow a crop on which communities depend (to give just one example of the many impacts climate change will have). There are other narratives that need to be told when it comes to climate and energy—narratives not only of environmental futures (though this too is important and will come into play) but also of the real character of the social and political present in relation to energy and the environment.

What connects the marvelous contributions to *Oil Fictions* is each essay's insistence on the need to expand our existing discussions of energy. It is tempting to describe this volume as an encounter between the physical and imaginative spaces, theories, and political concerns of the postcolonial and all that has come to constitute the Energy Humanities (however indefinite and imprecise this latter field might be). And it certainly *is* just this: a rich and varied interrogation of what an emphasis on energy reveals in the fictions of the Global South and, correspondingly, what those fictions add to what we (think we) know about energy. The essays in this volume explore fictions from

Africa, the Indian Ocean, the Middle East, Spanish America, and many more locales. They do so with an alertness to what Sheena Wilson has called "petro-intersectionality," a critical analytic that explores and exposes "how inequities of race, class, and gender are not only perpetuated in our current petroculture but also actively deployed as rhetorical strategies to literally and figuratively buoy and sustain existing power sources: oil and the neoliberal petro-state."[6] These essays describe the impact of energy on communities where the night sky is forever lit up by flares, invisible sour gas steals in to kill workers, and the lives of rivers are destroyed by chemicals and solvents—eco-traumas for which no one wants to take responsibility. They name, too, the specific experiences of Indigenous communities and of women, as both groups have borne much of the brute violence of extraction. It is impossible to come away from this collection without a sharper grasp not only of the scars that energy extraction has left on humanity and the earth but also of just how unequal the costs and benefits of petromodernity have been. The Global South and disenfranchised communities of the Global North have endured most of the consequences of petroculture, forced to contend directly with the ferocious battle over the planet's resources while enjoying few or none of the advantages that energy use affords.

This global division of power might seem obvious and elemental. And yet it is repeatedly lost in the hegemonic narratives of time imposed on the past, present, and future-still-to-come, which continue to insist that technological development and the expansion of liberal democratic practices (such as they are) constitute forms of "progress" we are supposed to collectively celebrate. The persistence of a widely held, quotidian faith in progress is remarkable: it remains alive even as antidemocratic populists assert their power around the world, from Hungary (a place to which I am no longer able to travel) to Brazil (where Bolsanaro has kicked "frontier capitalism" into high gear)[7] and innumerable places in between. A general faith in technology appears similarly difficult to shake, despite the cruelties and violence exerted by many forms of it, from the psychological scars left by internet trolling to the physical destruction carried out by military drones. *Oil Fictions* shakes to their core the self-certainties of the powerful fictions animating global liberal capitalism. If these essays just saw themselves as contributions to Postcolonial Studies and Energy Humanities by reading energy and resource extraction in the fictions of the Global South, they would be critically valuable and important. What makes them truly invaluable additions to the politics of the present (and, indeed, of the future, as highlighted here in the analysis of the unique fiction of Nnedi Okorafor) is how a focus on energy provides a renarration of history that might, finally, bring about a decisive end to the lie of modernity.

In her contribution, Swaralipi Nandi argues that there are deep and pervasive connections of the petro-present to the colonial past—links and ties that necessitate critical interrogation for what they reveal about both the histories and the futures of those communities that continue to endure the violence and dispossession of extraction. When critical emphasis is placed on extraction, we come to learn that the political victories that brought an end to colonialism, as momentous as these might have been, have rarely impacted ongoing operations of power, exploitation, and privilege, in part because they rarely impeded ongoing practices of resource extraction.[8] The history of colonialism and postcolonialism cannot be reduced to the voracious desire for control over resources. But this history and the continuation of all manner of brutal politics also cannot be fully told in the absence of the insatiable drive for control over resources. Property over life, power over people, resources over the environment, and violence and exclusions based on race and ethnicity: one comes away from reading *Oil Fictions* with a new sense of the history of modernity, one that redefines the modern as a reorganization of the mechanisms of power to ensure access to resources by the rich and powerful no matter who might be (officially) in charge of things. The impact on the people that inhabit the places of extraction and the labor of those who challenge imaginaries that treat them as disposable are the subjects of the fictions that the essays in this collection take up. To make the point that petro-economies continue the work of slave economies, as Nandi does, is to insist on the ongoing violence linked to resources and to point, too, to the exclusions and traumas that cannot help but emerge when the furnace of profits becomes the driving rationale of social life. This is no way to inhabit a planet, no way to live together.

The contributions offer powerful challenges to what they view as the assumptions and presumptions guiding the still-nascent field of Energy Humanities. The most important of these is the lack of attention in the Energy Humanities to the unique cultures of energy in the Global South. Yet by emphasizing the Global South and the ferocious battle over the planet's resources taking place there, it is inevitable that other things are missed in the process. This should come as no surprise—the critical focus on a missing ground (here, those "peripheral" modernisms in which energy is always already connected to violence, extraction, and dispossession) tends not to seamlessly complete the original figure (say, the role of energy in linking culture and power), thus giving us a complete picture of things, but to twist and bend it in ways that reveal yet other grounds that now need our critical attention. In short, *Oil Fictions* helps us to see not only the politics of energy in the Global South but also those questions we might yet still not be able or willing to ask ourselves about how we have been made—violently and unevenly—into creatures of energy. I make this point not to intimate that there are any problems with the current collection but to undertake what I understand the task of an afterword to be: to use the energies and insights of *Oil Fictions* to point to yet other ways that we might extend and expand our understanding of the mechanisms and processes that have shaped power, violence, and exploitation through the extraction of energy and other resources.

There are three such questions or issues that the book's varied, sharp, and expansive analyses brought to the fore for me. In her contribution, Helen Kapstein draws attention to the invisibility of oil in fiction and so, too, in much of literary criticism—so fully subsumed into experience that it is lost in the flow of everyday life. History has been shaped around rapacious practices of violent extraction; the power that comes with control over energy and other resources has meant that the bodies and habitats of plants, animals, and people have been trampled if they have stood in the way of powerful people who want access to the raw matter around which modernity has been shaped. Given the consequences of extraction—the consequences for the people of the Nigerian Delta, the Wet'suwet'en in British Columbia, and far too many other communities—how could one possibly suggest that oil is invisible (as Kapstein seems to do and as I and others have previously suggested as well)? Those whose lives are directly impacted by the violence and dirt of resource extraction have a different relationship to oil than do wealthy, first-world city dwellers, for whom electricity and gasoline appear as if by magic, stripped of the narratives of extraction that have made it possible to stick iPhone chargers into their walls with little thought as to where the power comes from. The essays in *Oil Fictions* tell us in no uncertain terms that, more than anything having to do with a resource's given character, whether oil is visible or invisible depends on who is encountering the resource and where they do so.

I think that this rejoinder to the use of the language of in/visibility, especially in relation to fossil fuels, is important. I also think that it sometimes misses what has been intended by those who have tried to speak of the unique character of oil vis-à-vis the landscape of modernity. Consider the careful language that Kapstein uses. She writes that oil "does not perform itself"; she claims that oil constitutes a "generative backdrop," something so much part of the daily stuff of life that it is hard to recognize oil's importance for how we live. The language of visibility draws attention to an important epistemic characteristic of oil, one that works much of the time, if not in cases in which the social and environmental impact of its extraction is obvious. What the essays in *Oil Fictions* point out is the need for a language that draws attention not to the *epistemological* operations of oil and energy so much as to their *ontological* characteristics. When Kapstein and others (including me) have spoken about the "generative backdrop" of oil that makes it invisible, the intent is not to deny that those who must endure the extractive violence of fossil fuels are intimately in contact with substances torn from the earth. Rather, it is to draw attention to how oil has enframed (*Gestell*) Being, in much the way that Martin Heidegger has argued modern technology has.[9] According to Heidegger, seeing modern technology as a means to an end, as we tend to do, misses how technology has ontologically reordered human existence—and in the deepest way imaginable. So it is, too, with oil, which has made itself seem to be the stuff that powers the technology of modernity—which it does—even while receding not from view (an epistemic claim) but from Being (an ontological one). The fundamental critical work of Energy Humanities isn't just to show how artists and writers have contended with the violence of extraction and the reality of petroculture. It is also to grapple with how oil has revised our ontological relationship to the Real, contributing to the idea of nature as little more than standing reserve and reshaping our ends to match its own.[10] Georges Bataille's distinction in *The Accursed Share* between a restricted and a general economy of energy makes a similar point.[11]

This leads me to a second issue. The texts explored in this volume constitute a significant expansion of the literature that addresses both the realities and imaginaries of oil. What is perhaps missing, however, is an interrogation of the processes and practices of literary criticism itself when it comes to oil and resources. All too quickly, it seems to me, energy has become yet another site or theme for the critical investigation of literature—a new subarea of research for those working in Environmental Humanities. This expansion of interest in oil has tended to sidestep the significant challenge that energy poses for the ways we undertake our analyses and, indeed, the reasons that we do so. Energy is essential to modern life; even so, it has tended not to be figured as such, made instead part of the background of life, part of the technological

Gestell described earlier. The texts assessed in this volume show that, despite this tendency, in certain situations and circumstances, and especially in those spaces and places still resistant to this *Gestell* or not yet enframed by it, the material and psychological drag and drain of resources poke through enough for writers to attend to the violence of fossil fuels. The present volume brings to the fore a new taxonomy of the literary and might constitute the first contribution to this taxonomy: there is extraction literature and literature that fails to contend with extraction, the latter being so immune to the realities of the earth's resources that it can be difficult to read this evasion even as a symptom (in a Jamesonian sense) or as an absence or suppression of some kind.

Even so, I feel it necessary to push my colleagues here: What exactly is it that we learn about the practices and processes of extraction from an assessment of *literature* about it? These essays and the literatures they interrogate reveal a different history of modernity, one in which human endeavor turns out to have always been something other than what it has represented to itself: not shiny progress or vibrant technology but raw matter used to exercise and maintain power, even if the majority of the planet's inhabitants have been ground up in the process. Given this new history, shouldn't we also rethink the how and the why of criticism as we have come to practice it? What does it accomplish to continue to attend to the literary using the rationales and orientations that emerged as part of this violent history of energy—to employ critical practices that cannot in any easy way be said to have developed outside of or in opposition to the history of extraction? Has literature been part of the problem of petrocultures more than we might imagine it to be—a device that, with exceptions, has worked to conceal our resource reality from us? What would it mean to refashion our critical approach to attend to the ontological power and capacities of energy—to make energy more than just a theme of our literary analysis but an animating principle of both its critical apparatus and its politics? These are questions I've been struggling with and have run up against in my own work; they form part of my assessment of energy periodization and its limits in my contribution to this volume. Our answers to these questions will reveal the need for a range of critical analyses for distinct forms of energy encounters while also giving us intriguing new insights into how energy has shaped (and continues to shape) the cultures we inhabit.

This brings me to the final issue to which I want to draw attention. In *The Great Derangement*, Amitav Ghosh argues that modern literature has been largely deaf to the environment.[12] For Ghosh, this has been the case right up to the present day. Despite the fact that we are now at the beginning of a climate catastrophe that will shape our experience for centuries (if not millennia), only a very small fraction of contemporary literary texts have bothered to contend with the environment (Ghosh exempts science fiction and poetry, without spending too much time explaining why). Ghosh claims that this lack of attention is due to the very character of the modern novel: its narrative, generic form has developed in a manner that makes it insensitive to the environment and allows it to contend poorly with climate change. At best, one might read the environment as the absent ground against which the rest of the work of the novel is carried out; at worst, the novel could be seen as a technological apparatus that has enframed its readers to view the planet and humanity's relationship to it in only very specific (and limited) ways. Ghosh poses the question of whether we look in vain to literature to animate the energies that will be needed to address the disaster already here and still to come, in this century and beyond. While the points Ghosh makes about the novel might be seen as far too sweeping, they can be taken as an invitation to give serious thought to the relationship of fiction and the environment—a litmus test, let's say, of the true politics of fiction in an era in which not addressing climate change constitutes something akin to a denial of the real circumstances in which we find ourselves today.

Climate and the environment are curiously absent from the essays in *Oil Fictions*. Their emphasis is on the human and social consequences of resource extraction, bringing to light the violent force that subtends modernity. The focus on colonial and postcolonial extraction is laudatory; as becomes all too clear, access to and control over resources are at the heart of modern power, violence, exploitation, and inhuman acts toward the human. This emphasis tends to put the environment to the side, it seems to me, if not push it to the background of critical concern. Is this yet another instance of the limit or absence pointed to by Ghosh when it comes to literature and the environment? An absence that can appear even in texts that emphasize oil and extraction? Might it be that we don't yet have the right conceptual lexicon or critical discourse to talk at the same time about the deformation of communities and clouds, selves and shells, history and planetary time? Let me be clear: I'm not suggesting that there isn't any attention to the environment in these pages or that this gap constitutes a failure or limit of some kind. The point is simply to draw attention to a propensity that subtends these contributions so that we might engage in forms of criticism alive to the human *and* nonhuman in each and every moment of narrative analysis.

The rich and varied contributions to *Oil Fictions* constitute a major intervention into our analysis of energy and culture. The powerful and insightful introduction by Stacey Balkan and Swaralipi Nandi points out what is at stake in our critical labors while giving us many of the analytic resources needed to answer the questions I've posed here. Together, the pieces gathered in this collection not only give us insights into our petrofictions but also offer us tools to formulate new questions and to begin the task of answering them. By doing so, we might yet be able to fashion worlds and imaginaries beyond the violent and exclusionary ones we inhabit today.

Notes

1. For a range of recent statistics on worldwide energy consumption and production, see the June 2018 edition of the *BP Statistical Review of World Energy*, available at https://www.bp.com/content/dam/bp/en/corporate/pdf/energy-economics/statistical-review/bp-stats-review-2018-full-report.pdf.

2. See Dahr Jamail, "When the Ice Melts: The Catastrophe of Vanishing Glaciers," *The Guardian*, January 8, 2019, https://www.theguardian.com/news/2019/jan/08/when-the-ice-melts-the-catastrophe-of-vanishing-glaciers.

3. See Leyland Cecco, "Canada Spawns Its Own Yellow Vest Protests—with Extra Rightwing Populism," *The Guardian*, December 20, 2018, https://www.theguardian.com/world/2018/dec/20/canada-yellow-vest-protests-gilets-jaunes.

4. Leyland Cecco, "14 Arrested at Indigenous Anti-pipeline Protest Camp as Tensions Rise," *The Guardian*, January 8, 2019, https://www.theguardian.com/world/2019/jan/08/canada-14-arrested-indigenous-anti-pipeline-protest-camp.

5. Karina Baptista offers a broad and useful overview of the term "petroculture" in her contribution to "Global South Studies: A Collective Publication." See https://globalsouthstudies.as.virginia.edu/key-concepts/petrocultures.

6. Sheena Wilson, "Gender," in *Fueling Culture: 101 Words for Energy and Environment*, ed. Imre Szeman, Jennifer Wenzel, and Patricia Yaeger (New York: Fordham University Press, 2017), 177.

7. See Daniel Cunha, "Bolsonarism and 'Frontier Capitalism,'" *Brooklyn Rail*, February 5, 2019, https://brooklynrail.org/2019/02/field-notes/Bolsonarism-and-Frontier-Capitalism.

8. There are, of course, exceptions to this general trend, though they are few and far between. Thea Riofrancos has examined how even apparently left-wing governments, such as that of Rafael Correa in Ecuador, have continued to engage in extraction in a manner that aligns left-state practice with private, neoliberal approaches to the environment as a resource for economic development. See Riofrancos, "*Extractivismo* Unearthed: A Genealogy of a Radical Discourse," *Cultural Studies* 31, nos. 2–3 (2017): 277–306.

9. Heidegger's assessment of the ontological significance of technology can be found in "The Question Concerning Technology,"

in *The Question Concerning Technology and Other Essays*, trans. William Lovitt (New York: Harper and Row, 1977), 3–35.

10. Reza Negarestani's one-of-a-kind theory-fiction *Cyclonopedia: Complicity with Anonymous Materials* explores the ontological singularity of oil and its sociopolitical significance better than almost any other existing text. For Negarestani, oil is a "satanic sentience" nurtured by the planet's petropolitics, whose true power is that no one on earth knows that they are engaged in prayers and sacrifices to a demon that thrives on war and conflict. See Negarestani, *Cyclonopedia: Complicity with Anonymous Materials* (Melbourne: re.press, 2008).

11. Georges Bataille, *The Accursed Share*, trans. Robert Hurley, vol. 1 (Cambridge, MA: Zone Books, 1987).

12. Amitav Ghosh, *The Great Derangement: Climate Change and the Unthinkable* (Chicago: University of Chicago Press, 2017).

Contributors

Henry Obi Ajumeze received a PhD from the Centre for African Studies, University of Cape Town, specializing in the interdisciplinary connection between drama, ecology, and petroculture. He teaches dramatic theory and criticism at Bowen University in Nigeria. He is a fellow of the African Humanities Program of the American Council of Learned Societies, and he is currently working on a book project titled *The Dialogue Between Oil and Water in Nigeria's Niger Delta.*

Rebecca Babcock is the William and Ordelle Watts Professor at the University of Texas Permian Basin, where she teaches courses in writing and linguistics. She also serves as the Freshman English Coordinator and Director of Undergraduate Research. She has authored, coauthored, or edited several books on tutoring, writing centers, disability, and metaresearch. Her latest book is *Theories and Methods of Writing Center Research*, edited with Jo Mackiewicz. She has also published research articles in *Writing Lab Newsletter*, *Linguistics and Education*, *Composition Forum*, *Praxis*, and *The Peer Review.*

Stacey Balkan is assistant professor of environmental literature and humanities at Florida Atlantic University, where she also serves as an affiliate faculty member for FAU's Peace, Justice, and Human Rights Initiative. She has published several works on energy cultures including "Rogues in the Postcolony: Chris Abani's *GraceLand* and the Petro-picaresque" (*Global South*, 2015), "Energo-poetics: Reading Energy in the Ages of Wood, Oil, and Wind" (*Études anglaises*, 2021), and her forthcoming monograph *Rogues in the Postcolony: Narrating Extraction and Itinerancy in India* (West Virginia University Press). Stacey's recent work also appears in *ISLE: Interdisciplinary Studies in Literature and Environment*, *Global South Studies*, *Mediations*, *Social Text Online*, and *Public Books.*

Ashley Dawson is professor of postcolonial studies in the English Department at the Graduate Center / City University of New York and the College of Staten Island. His most recent books include *People's Power: Reclaiming the Energy Commons* (O/R, 2020), *Extreme Cities: The Peril and Promise of Urban Life in the Age of Climate Change* (Verso, 2017), and *Extinction: A Radical History* (O/R, 2016). A member of the Social Text Collective and the founder of the CUNY Climate Action Lab, he is a longtime climate justice activist.

Sharae Deckard is lecturer in world literature at University College Dublin. She has published multiple articles on world-ecology and energy humanities approaches to world literature, and four books, the most recent of which is an edited collection, *World Literature, Neoliberalism, and the Culture of Discontent* (Palgrave, 2019). She has also edited several journal issues on the topic, including an issue of the *Irish University Review* on "Food, Energy, Climate: Irish Culture and World-Ecology" and an issue of *Green Letters* on "Global and Postcolonial Ecologies." New essays on world-literary crime fiction and fossil capital and on the intersection of oil and hydrodependency are forthcoming in *Études anglaises* and *New Formations.*

Scott DeVries received a PhD in Spanish from Rutgers University in 2004 with a dissertation on ecology and environmentalism in recent Spanish American fiction. Currently he works as an independent scholar of literature, energy, and ecology. His continuing research encompasses both nineteenth- and twentieth-century Latin American literature from the perspective of ecological criticism, animal studies, and energy. His first book, *A History of Ecology and Environmentalism in Spanish American Literature*, published by Bucknell University Press in 2013, and his second, *Creature Discomfort: Fauna-Criticism, Ethics, and the Representation of Animals in Spanish American Fiction and Poetry* was published by Brill in 2016.

Kristen Figgins is a doctoral candidate in English at the University of Arkansas. Her specialization is nineteenth-century British literature, critical animal studies, and adaptation. Her current research involves tracing how developments in natural science and animal rights philosophy are adapted in transhistorical literature. Her recent research can be found in *Literature/Film Quarterly* and in her co-edited book *Boom or Bust: Narrative, Life, and Culture from the West Texas Oil Patch* (OU Press).

Amitav Ghosh was born in Calcutta and grew up in India, Bangladesh, and Sri Lanka. He is the author of two books of nonfiction, a collection of essays, and ten novels. His books have won many prizes, and he holds four honorary doctorates. His work has been translated into more than thirty languages, and he has served on the jury of the Locarno and Venice film festivals. In 2018 he became the first English-language writer to receive India's highest literary honor, the Jnanpith Award. His most recent publication is *Gun Island*, a novel.

Corbin Hiday is a PhD candidate in English at the University of Illinois at Chicago. His dissertation explores how a series of novels from different historical moments of the British Empire theorize the constitutive entanglement between progress narratives and the reality of resource exhaustion, a contradiction that continues to drive contemporary fossil-fueled catastrophes. His publications have appeared in *The Bloomsbury Companion to Marx*, *b2o: An Online Journal*, and the V21 Collations Book Forum.

Helen Kapstein is associate professor in the English Department at John Jay College, The City University of New York. She earned her PhD in English and comparative literature from Columbia University. Her book *Postcolonial Nations, Islands, and Tourism: Reading Real and Imagined Spaces* was published in 2017 (in paperback in 2019) by Rowman and Littlefield International. Her work, including an essay titled "Crude Fictions," on how Nigerian short stories sabotage Big Oil's narrative, has appeared in *Postcolonial Text*, *English Studies in Canada*, and *Safundi: The Journal of South African and American Studies*, among other venues.

Swaralipi Nandi is an assistant professor of English at Loyola Academy, Hyderabad. She has a PhD in English from Kent State University with a research focus on postcolonial literature in the era of neoliberal globalization, and she is currently working on postcolonial environmental studies. Her work has appeared in *Interventions: A Journal of Postcolonial Studies*, *Journal of Narrative Theory*, *Studies in Travel Writing*, *Literary Geographies*, *JSL: The Journal of the School of Language, Literature & Culture Studies*, *Worldly Teaching: Critical Pedagogy*, and *Global Literature*, among others. She has also edited two books: *The Postnational Fantasy: Essays on Postcolonialism, Cosmopolitics, and Science Fiction* (McFarland) and *Spectacles of Blood: A Study of Violence and Masculinity in Postcolonial Films* (University of Chicago and Zubaan)—and her forthcoming monograph, *Narrating the Fringes: The Changing Narrative of Poverty in 21st Century Indian English Fictions*, is to be published by Routledge.

Micheal Angelo Rumore is a doctoral candidate in English at the Graduate Center, CUNY. His dissertation, titled "Toward the Black Indian Ocean: Race and the Human Project in the Afro-Asian Imagination," interrogates why dominant notions of oceanic "cosmopolitanism" appearing frequently in the field of Indian Ocean studies tend to exclude Blackness and Africanness. His writing has also appeared in *Social Text Online*, *Studies in the Fantastic*, and *Guernica*. In addition, he teaches literature and writing courses at Lehman College, CUNY.

Simon Ryle is an associate professor at the University of Split, Croatia. His research focuses on intersections of ecopoetics, creaturely humanities, and literary infrastructures. He has recently edited an issue of the *Journal for Cultural Research* on the topic of "Minor Shakespeares" and is currently coediting an edition of *Cross-Cultural Studies Review* titled "Wavescapes in the Capitalocene." His first monograph was published by Palgrave Macmillan. His research has also appeared in various volumes and journals, including *boundary 2* (Duke University Press), *Umjetnost riječi* (Hrvatsko filološko društvo), *Cahiers Élisabéthains* (Centre national de la recherche scientifique), *Adaptation*

(Oxford University Press), *Shakespeare Survey* (Cambridge University Press), and *Textual Practice* (Routledge).

Sheena Stief is currently employed at the University of Texas Permian Basin. She teaches Freshman Composition as well as British and American Literature post-survey courses. Her research centers around dual credit studies in the online composition classroom as well as environmental aspects in the genre of paranormal romance. She graduated from the University of Texas Permian Basin in 2013 with a master's in English. She is also a freelance editor. Sheena resides in Texas with her husband of thirteen years, three children, and three dogs. She can be found online through her website, www.stiefediting.com, and her Twitter handle, @sheena_stief.

Imre Szeman is University Research Chair in Communication Arts at the University of Waterloo, where he is also a member of the Interdisciplinary Centre on Climate Change. His recent publications include *On Petrocultures: Globalization, Culture, and Energy* (2019) and *Energy Culture: Art and Theory on Oil and Beyond* (coedited with Jeff Diamanti, 2019). He is currently working on *Energy Impasse: On the Limits of Transition* (with Mark Simpson).

Maya Vinai works as an assistant professor in BITS Pilani (Hyderabad campus). Her research interests include maritime narratives along the coastal belt of Malabar, contemporary Indian English fiction, and temple art forms of South India. Along with several publications in both national and international journals, she has authored a book titled *Caste and Gender in the Works of Anita Nair.*

Wendy W. Walters is a professor in the Department of Writing, Literature, and Publishing and director of the honors program at Emerson College, Boston. She is the author of two books, *Archives of the Black Atlantic: Reading Between Literature and History* (Routledge, 2013) and *At Home in Diaspora: Black International Literature* (University of Minnesota Press, 2005). She has published articles in *Callaloo, African American Review, American Literature*, and elsewhere. She is currently working on a book about African diasporic ecocritical literatures and has been teaching a course on Afrofuturism since 2004.

Index

States
Publisher Services